Fundamentals of
Plant-Pest Control

Fundamentals of Plant-Pest Control

Daniel Altman Roberts
University of Florida

with chapters contributed by
Robert E. Stall and Daniel A. Roberts
Vernon G. Perry
Sidney L. Poe
Earl G. Rodgers
Jerome V. Shireman
all of the University of Florida

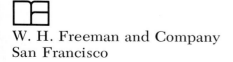
W. H. Freeman and Company
San Francisco

Library of Congress Cataloging in Publication Data

Roberts, Daniel A.
    Fundamentals of plant-pest control.

Cover: A woodcut illustration of a bean plant ("Phaseolus peregrinus") from *Rariorum Plantarum Historia,* an herbal by Carolus Clusius, published in Antwerp in 1601. [From *Plant and Floral Woodcuts for Designers and Craftsmen,* selected and arranged by Theodore Menten. Dover Publications, 1974.]

    Includes bibliographies and index.
    1.   Pest control.   2.   Agricultural pests.   3.   Plant
diseases.   I.   Stall, Robert E.   II.   Title.
SB950.R6   632'.9   77-16135
ISBN 0-7167-0041-7

Fundamentals of Plant-Pest Control

Printed in the United States of America

9 8 7 6 5 4 3 2 1

to Ruth Remsen Roberts

Contents

Preface

Noxious organisms, collectively termed pests, have plagued mankind from the beginning, and, it seems, will continue to vex people and thwart their endeavors to the end. The pests that beset human beings and other animals and that threaten green plants abound in households, cities, barns, storage bins, fields, and the wilderness. Pests must be kept under control if we are to feed, clothe, and shelter ourselves. And by control I mean keeping pest populations at or below their economic thresholds, those elusive levels at which the cost of combative efforts equals the value of gains achieved by those efforts. This book treats, at the first level of understanding, the nature and control of only those pests that attack crop plants, which are plants that people manage, at least to some extent.

The book has three parts. The first part treats plant-pest control as a science that unifies the several disciplines concerned with plant pests and their control. The second part introduces the several disciplines and gives an overview of procedures used in the control of plant pathogens (including nematodes), harmful insects and other arthropods, weeds, and vermin. The third part considers the systems approach to plant-pest control in an agricultural perspective. The book is meant to make students of schools, colleges, and universities aware of plant-pest control as a unifying multidisciplinary science in its own right. It is also meant to provide a foundation on which plant-pest-control specialists of the future may build useful and productive careers.

Five of my colleagues have especially helped this book fulfill its purposes; I am greatly indebted to them for contributing the authoritative chapters that make up Part II. And all of us are grateful to Charles B. Browning, Dean of the College of Agriculture, Institute of Food and Agricultural Sciences, University of Florida, Gainesville, for fostering our efforts and for establishing in our college a special program of undergraduate education in systems of pest control for plant protection. We are also grateful to our departmental and unit chairmen, L. H. Purdy (Plant Pathology), W. G. Eden and F. G. Maxwell (Entomology and Nematology), C. Y. Ward (Agronomy), and J. L. Gray (Forest Resources and Conservation), with whom administrative matters were handled.

I am especially grateful to my dear wife, Ruth Remsen Roberts, who carried out the difficult task of typing the revised manuscript, not only swiftly and accurately, but cheerfully.

Finally, this book is offered to those embarking on their study of plant-pest control as a science. We hope it serves them well.

January 1978   *Daniel Altman Roberts*
Gainesville, Florida

Fundamentals of
Plant-Pest Control

I
THE SCIENCE OF
PLANT-PEST
CONTROL

1

Agriculture, Plant Pests, and Ecosystems

Earth is the planet of green plants. Plants are the verdant mantle covering much of the Earth's land; they are the living turquoise masses flourishing in the shallows of the seven seas. Green plants, by the miracle of photosynthesis, capture a tiny fraction of the Sun's awesome physical energy and convert it into enough chemical energy to support all life on this planet. Without green plants, all living things would perish, and the Earth would revert to its original barren state. Mankind survives on Earth only because green plants grow here. Indeed, "all flesh is grass" (Isaiah 40:6).

AGRICULTURE

For eons—perhaps as long as three million years—our forebears took what little nourishment they could find in the forests, oceans, inland waters, and plains. They hunted animals, they fished, and they gathered greens, berries, grain, nuts, roots, and other plant products. They consumed their harvests raw at first; then cooked after control of fire was discovered about 500,000 years ago. Domestication of animals and cultivation of plants for food are relatively recent innovations, dating from only 7,000 to 10,000 years ago.

Agriculture thus dates from the Neolithic Age. It may have originated in the hills flanking the "Fertile Crescent" around greater

Mesopotamia (Figure 1.1), from where it could have radiated in all directions, developing with time, until all the inhabited world was engaged in crop production and animal husbandry. Or, agriculture may have originated and developed independently in several such centers (Asia, Africa, Europe, the Americas), and then spread its technology in ever-widening and overlapping circles. Whichever is true, agriculture, with fire, became the very foundation of civilization. Fully 10% of the Earth's land mass is now devoted to producing crop plants to feed us; about 20 percent is used for forage and pasture plants to feed our animals, and almost 30 percent is cloaked in forest trees, which give us much of our shelter. Areas of arable land are shown in Figure 1.2. The remaining 40 percent of the Earth's land consists of mountains, deserts, and frozen tundra, except for cities and highways, which seem to ooze relentlessly over the land, at least in highly industrialized countries.

There are approximately 200,000 species of seed plants, (Fuller and Ritchie, 1967), but crop plants—meaning those seed plants that are managed, at least to some degree, by human beings—number only in the tens of thousands. Of these, only about a thousand different species play a part in our economic activities, and hardly more than a hundred have any financial significance in international markets. But most astonishing: only 15 species of plants and nine species of animals provide almost all the food for the world's population (Janick et al., 1974; Ehrlich and Ehrlich, 1970). The 15 staple food plants (rice, wheat, maize, barley, sorghum, sugar cane, sugar beet, potato, sweet potato, cassava, bean, peanut, soybean, coconut, and banana), the forage grasses and legumes, the fiber and forest plants, and the nine principal domestic animals (cattle, swine, sheep, goats, water buffaloes, chickens, ducks, geese, and turkeys) have for centuries received the almost undivided attention of most agriculturists, and properly so.

This book treats the subject of green plants—those named above and others, such as additional vegetables, fruit and nut plants, and ornamental herbs and shrubs—but only as those green plants are victimized by competitive, parasitic, and injurious organisms, including herbivores. This book thus concerns crop-plant pests and their control.

PLANT PESTS

Crop-plant production is fraught with hazards. Plants may be inundated by floods and tattered by hailstones; they may parch and burn

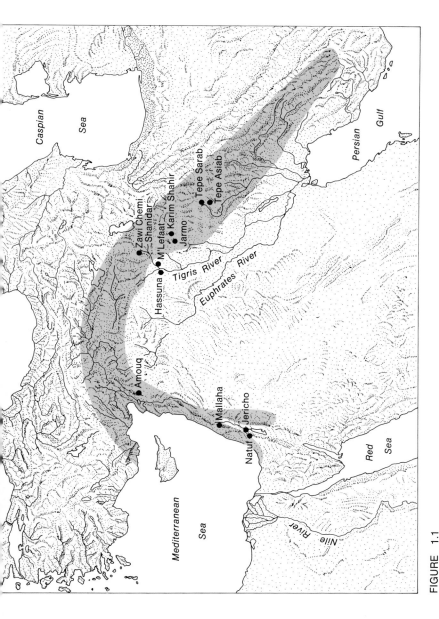

FIGURE 1.1

The shaded area shows the grassy and forested hills, whose soils and climate favored plants (wild wheat, barley, and other food plants) and animals (goats, sheep, swine, and cattle). Farming was practiced in the Fertile Crescent at the foot of these hills approximately 10,000 years ago. The locations of early permanent settlements that have been excavated are shown. [Adapted from "The agricultural revolution" by Robert J. Braidwood. Copyright © 1960 by Scientific American, Inc. All rights reserved.]

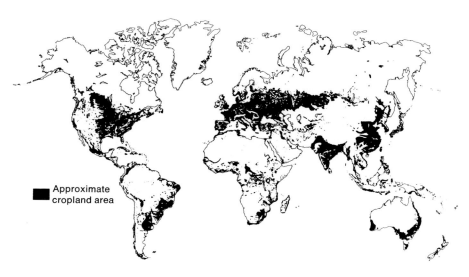

FIGURE   1.2

Approximate arable land areas of the world are shaded. [From USDA, miscellaneous publication 705, 1964.]

during droughts; they may blacken and die by the thunderbolt. And we are all but helpless in the face of such natural disasters. The forests and the savannas have yet another enemy, fire, which consumes, with incredible swiftness, everything in its path. We have not yet achieved much success in preventing forest and grass fires kindled by our own kind; we certainly cannot prevent those sparked by lightning.

All crop plants face a final hazard, that of their natural plant and animal enemies, plant pests.

A Definition of Plant Pests

Plant pests are organisms noxious to plants. They are living things, or, as with plant viruses, they behave like living things. Plant pests are harmful to the functioning of green plants, whose form those pests usually distort. Weeds compete with crop plants; pests that cause disease in plants continuously disrupt plant functions; all other pests disrupt plant functions discontinuously, thus causing injury rather than disease (Roberts and Boothroyd, 1972). Whether suffering from competition, disease, or injury, the victimized plant usually dies before its time; it always yields less than full measure.

Perhaps in the broadest sense of the term, pests should be defined as pernicious organisms, in recognition of their capacities to make mischief or to vex, as well as to do harm to their victims. That organism which a pest vexes or harms is termed the victim (Bierne, 1967). The host is thus the victim of the parasite, the prey the victim of the predator, and so on. The couplet pest-victim, therefore, implies the multifarious interactions that the two members engage in, and that are always contrary to the physiological and anatomical well-being of the victim.

Because the well-being of crop plants governs the welfare of those who practice agriculture and those who reap its benefits, some choose to include the notion of harmfulness to humans in their definitions of crop-plant pests. Pests have been defined, for example, as "organisms inimical to the welfare of man" (Tschirley, 1972); "harmful organisms" from the human point of view (Bunting, 1972); and "any species that is unacceptably abundant" (Bisplinghoff and Brooks, 1972), presumably, unacceptably so to human beings. Recognizing that crop-plant pests do indeed vex people and thwart their plans, I prefer, nevertheless, to assign the central role to the victim, the direct object of the pest's attack—in this book, the crop plant itself.

Crop Losses Due to Plant Pests

That plant pests inflict losses on their victims is implicit in the definition of pests as noxious organisms. Also, the destructiveness of plant pests is patently reflected in the reduced vigor and yield of crop plants and in the debased quality of their products. But no one has yet ascertained, with any degree of accuracy, the magnitude of crop losses due to pests, whether such losses have been expressed in terms of decreased productivity, percentage losses after harvest, presumed monetary values of reduced yields, or costs of pest-control methods. All we have are estimates, one about as accurate—or inaccurate—as another.

But a conservative consensus seems to have been reached: plant pests may well destroy, world-wide, fully one-third of humanity's food and fiber supply from plants every year (Janick et al., 1974). And the consensus figures may err on the low side, for Beirne estimates:

> The average real total material (as distinct from monetary) losses of all kinds—observed, unnoticed, and invisible—caused annually by all kinds of pests—insects, disease organisms, nematodes, vertebrates, weeds—may perhaps be somewhere between 25 and 50 per cent in regions such as North America and Western Europe, where control measures are applied relatively intensively, and up to 80 per cent or more elsewhere. These are guesses, but as reliable or unreliable as any other similar figures.

Beirne's guesses could be right. For example, $7 billion were lost to plant pests and over $3 billion were spent on pest-control efforts annually in the United States during the 1960's (Janick et al., 1974). Plant pests, therefore, cost—in control procedures and in uncontrolled losses combined—about one-fourth the value of crop plants in the country. But such a loss is nothing at all compared to the tragic losses of foodstuffs in undeveloped countries. India probably loses more than 10 percent of her grain every year to a single vertebrate pest, the rat; birds are said to destroy $7 million worth of crop plants a year in Africa; in two provinces of the Philippines, rats devoured 90 percent of the rice, 20 to 80 percent of the corn (maize), and more than 50 percent of the sugar cane in a single year. Insects alone destroy up to 50 percent of the stored grain in some years in many undeveloped countries, where, on the average, grain losses *after* the harvest to insects, vermin, and spoilage doubtless amount to 20 to 80 percent annually (Ehrlich and Ehrlich, 1970). And so it goes, on and on and on.

Losses attributed to crop-plant pests have been reduced over the years in technologically developed countries, but only up to a point. For decades, it seems that farmers in the United States have lost about 10 percent of their crops in the field to insects. Early reports (Metcalf and Flint, 1932) suggest this, but so do reports in 1948 and 1969 (Ehrlich and Ehrlich, 1970). Figures on losses to plant diseases in the United States are much the same. For example, estimated losses of wheat to plant diseases averaged a little more than 10 percent between 1919 and 1937 (Chester, 1947), slightly more than that between 1951 and 1960 (LeClerg, 1964).

Have we in the United States gone about as far as we can go toward controlling crop-plant pests? I think not, for we can undoubtedly reduce our pest-control costs, both cost in money and cost in damage to the environment. Moreover, plant-pest control remains woefully inadequate in undeveloped countries. We need to redouble our efforts against plant pests; it would be foolish in the extreme to assume an attitude of either complacency or defeatism toward controlling pests of the plants that feed, clothe, and shelter human beings and their animals.

The continual decimation of crop plants by insects, diseases, weeds, or vertebrate pests can be explained on the assumption that, as one pest is controlled, another takes its place. The solution to one problem may thus invite disaster from a different and even more damaging source. As we improve farm-management practices, replace old-line varieties of crop plants with new ones, and practice modern pest-control methods, we present plant pests with drastically altered environments. Consequently, populations of plant pests change as those species adapt to their new environments. As a once-dreaded pest virtually disappears, another, perhaps previously unnoticed, may burgeon forth and destroy the crop. For example, rust-resistant oat varieties of "Victoria" parentage from Australia replaced the common varieties on Iowa farms in the 1940's. But the new rust-proof varieties soon succumbed to another malady, Victoria blight, caused by the parasitic fungal pest, *Helminthosporium victoriae,* endemic to the United States. The solution to one problem, rust, thus created a new problem, blight. The point is that agriculturists, if they would combat crop-plant pests effectively, must gird for a war of continued attrition against those pests. The struggle will surely continue as long as mankind and agriculture exist, for we will never eliminate any plant pest from the face of the Earth.

An Enumeration of Crop-Plant Pests

The reliability of estimates of losses, discussed in the foregoing section, may well be questioned. Were pest-loss surveys made and, if so, were they made in truly representative regions? Could the visible damage done by pests be positively and accurately related to decreased productivity of crop plants? Did significant damage go undetected? What were the relationships between estimated production losses and presumed monetary losses? However questionable average crop-loss figures may be, the individual farmer knows full well when he loses all or part of his cash or food crop during a given growing season. He learns the nature and extent of crop losses to plant pests from bitter experience.

Any contributions that plant-pest-control programs might make to world agriculture and thus to helping solve the problem of world hunger will be made, in the final analysis, by the knowledgeable farmer himself. What are the pests to be contended with by the corn farmer in the United States, the wheat farmer in Russia, the potato farmer of western Europe, the rice farmer in China? Plant pests include weeds, vermin, insects and other arthropods, and plant pathogens (Figure 1.3).

Weeds

Higher plants that compete with crop plants for space, water, and nutrients—and that often harbor crop-attacking insects and parasites—are all weeds. Weeds are plants despised by farmers, but green plants qualify as weed pests of crop plants only when they reduce the growth and yield, and impair the quality, of the plants that people manage (Chapter 9).

FIGURE   1.3

The principal groups of plant pests are insects (corn earworm, a), plant parasites (corn smut, b), vermin (rat, c), and weeds (common lambsquarters, d).

Vermin

I regard all vertebrate pests as vermin. Those that affect crop plants and their products include such herbivores as rabbits, certain birds, deer (on occasion), and other wild creatures. But the moxt noxious of all vermin is the rat (Chapter 10). This loathsome animal not only robs human beings of their sustenance and gnaws away the very foundations of their shelter, but plays the role of a deadly courier of pestilence. In highly industrialized societies, the rat has pretty much abandoned the farm for the blighted inner cities, but the rat remains perhaps the number-one enemy of farmers in undeveloped countries. There, this dangerous rodent first ravages the growing crop, and then takes a major portion of the harvest that has been stored against hunger.

Arthropods

Not all the classes of arthropods (which include the shrimp, lobster, and other crustaceans) contain species of plant pests, but many insects and a few arachnids (mites, for example) are among the most devastating of all plant pests (Chapter 8).

Plant Pathogens

The infectious (that is, transmissible) agents that cause disease in plants are termed plant pathogens. They include certain viroids, viruses, mycoplasmalike organisms, rickettsiae, bacteria, fungi, algae, parasitic seed plants, protozoans, and plant-parasitic nematodes. These organisms, so diverse in form, have one common characteristic: they all cause disease in green plants. These pests are treated in Chapters 6 and 7.

Literally hundreds of thousands of species of living things are represented in the principal groups of plant pests just enumerated. Populations of these noxious organisms make up a significant fraction of the life on planet Earth. The agriculturist must reckon with plant pests, then, as integral parts of agricultural and natural ecosystems.

ECOSYSTEMS

Human beings have seldom foreseen the impact on their environment of the technological innovations that they have proudly put into practice on their farms. This lack of foresight has prompted Ehrlich and Ehrlich (1970) to declare world agriculture an "ecological disaster area." Such an indictment should give us pause. Agriculturists must unanimously view agriculture as part of the Earth's ecosystem. And plant-pest-control specialists, because they have contributed to the ecological problems spawned by the practice of modern agriculture, owe it to themselves and their constituents to conduct their activities with the ecological ramifications of pest-control programs uppermost in their minds. We therefore need to be familiar with the language and the basic concepts of ecology, "the study of ecosystems" (Odum, 1971).

Terminology

Terminology, the correct use of the names that are given to concepts, species, structures, and phenomena (Whetzel, 1929), is the cornerstone of scientific communication. Many of the controversies that occupy the pages of scientific periodicals originate in misunderstandings over the meanings of words. A few of the ecological terms that are important to the student of plant-pest control are defined in this section.

Taxa

The discrete and identifiable units into which living things are classified have been termed *taxa;* the singular is *taxon.* The so-called major taxa include, in decreasing order of diversity, kingdoms, divisions (phyla), classes, orders, and families; genera and species are the minor taxa (Janick et al., 1974). These groups, regardless of their position in the scheme, are described principally in terms of morphological features—that is, form, rather than function—particularly those of the sexual reproductive parts. Each group may, on occasion, be further subdivided, and, except for certain subdivisions of the species, these subunits are also separated from one another on the basis of form.

The species, the fundamental unit of taxonomy, is a stable and distinct population of interbreeding forms of life characterized by similar morphology. Together with the next higher-ranking taxon, the genus, the species is the taxon on which we base nomenclature, the scientific naming of organisms. The scientific name of an organism always consists of two terms, the name of the genus and the name of the species, both in Latin.

This book is primarily concerned with species of crop plants and with species of their pests. But certain subdivisions of species are also important. For example, agricultural *varieties* within crop-plant species play significant roles in agriculture, and, if pest-resistant, in plant-pest control. Each such variety, unlike botanical subdivisions of a species (subspecies, botanical varieties), is distinguished by its unique horticultural qualities, not by its morphological features. The agricultural variety, often termed the cultivated variety (cultivar for short), may be: clonal, as in plants propagated vegetatively; a pure line, in self-pollinated species; an open-pollinated variety, in plants that are cross-pollinated; a synthetic variety, in progenies of a cross-pollinated species growing in polycross nurseries; or a hybrid variety, when progenies of two or more inbred lines are selected as the variety. When used alone, without a modifying adjective, the term variety, at least in agricultural literature, almost always refers to one of the several kinds of agricultural varieties.

Important subdivisions of plant-pest species will be discussed in subsequent chapters. Such subdivisions, like the varietal subdivisions of crop-plant species, are generally distinguished not by morphology, but by physiological or pathological characteristics and capabilities.

Populations

In its original sense, population meant simply the number of people inhabiting a geographical or political region. We say, for example: the population of the Earth is approximately four billion. Biologists, however, have borrowed the term and extended its meaning to denote "groups of organisms having common origin, form, and function" (Boughey, 1971).

Communities

Groups of populations (not morphologically related, but related to each other by interdependent functions) are termed communities (Boughey, 1971).

Ecosystems

"Conceptual systems . . .within which communities exchange energy, materials, and information with one another and with their physical environment" are termed ecosystems (Boughey, 1971). The ecosystem concept is the basis of all ecology.

Ecology

Often referred to as "the science of survival," ecology is "the study of communities in relation to their environment" (Boughey, 1971). Ecology thus integrates the concepts of species, populations, communities, and ecosystems. Study of the biomass—the living populations of plants and animals in an ecosystem—is called synecology; the study of the individual and its species, autecology. The specialist in the control of plant pests is concerned not only with the form and function of individual species of plant victims and their pests, but also with all their interactions as they influence ecosystems and as they are influenced by other components, living and nonliving, of their ecosystems.

Shelford's Law of Toleration

The nonliving (abiotic) elements of an ecosystem profoundly affect the functioning of the living elements therein. In fact, the performance of a plant or an animal, the very integrity of a population or community, may depend entirely on a single abiotic factor, such as temperature or moisture. For every environmental factor that influences a population of living things, there is an optimum range within which the organism will flourish, and minimum and maximum ranges within which the population will function poorly, if it functions at all. The limits of tolerance (Figure 1.4) are the lowest and highest measured values of the abiotic factor within which a given biotic element can function (Shelford, 1913).

The concept of the law of toleration can be broadened to include interactions between two living components of an ecosystem, as between victim and pest. But only half the concept applies, that half relating to maximum values. For instance, the population of a victim species will remain abundant and its members will function well only so long as the population of a pest species remains below some maximum level. Once the pest population reaches or surpasses this level of abundance, however, the population of the victim species will decline, or its individuals will suffer impaired performance, because of competition, predation, or parasitism.

Blackman's Limiting-Factor Concept

The concept of limiting factors was proposed by Blackman (1905) to clarify the interactions between temperature, light intensity, and carbon dioxide as factors that influence the rate of the process of photosynthesis (Figure 1.5). The limiting-factor concept originated, however, in Liebig's law of the minimum, which stated that a soil would not permit good growth of plants if it had too little of even one of the essential mineral nutrients. Crop yields would, therefore, depend on the plant nutrient that is present in the least, or minimum, amount. Several interdependent and interacting factors determine the rate of any complex biological process, but not all these factors control the rate at a given time. Instead, only one deficient factor might

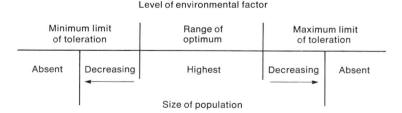

FIGURE   1.4

Diagram of how the law of toleration applies to the distribution and population of a species. [From *Animal Communities in Temperate America,* by V. E. Shelford. Published by the University of Chicago Press. Copyright © 1913 by the Geographic Society of Chicago. All rights reserved.]

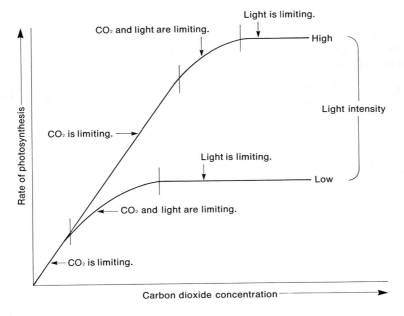

FIGURE   1.5

The limiting-factor concept in photosynthesis (Blackman, 1905). Carbon dioxide first limits the rate of photosynthesis because the rate of the process increases proportionately with the increased concentration of carbon dioxide. Finally, light becomes limiting, and the rate of photosynthesis will not be affected at all by increased concentrations of carbon dioxide. Just before light alone becomes limiting, the rate of photosynthesis is limited by carbon dioxide as well as by light.

determine the rate of a process at first; then, if that deficiency is corrected, another factor might become deficient later, and so on. A limiting or deficiency factor, then, is "any essential factor which is deficient in a degree such that an increase in its amount will directly increase the rate of the process" (Curtis and Clark, 1950).

The Retarding-Factor Concept

Some authors try to class any factor that influences the rate of a biological process as a limiting factor. But such imprecision in the use of terms masks the nature of biological phenomena and leads to confusion and controversy. What kind of factor is a pesticide, which by its presence or abundance—not by its absence or "deficiency"— adversely influences the rate of growth in a population of pests? A pesticide certainly is not a limiting factor, at least, not in the correct sense of the term. Rather, a pesticide is a retarding factor, one that inhibits. Curtis and Clark (1950) used the term retarding factor to designate any factor that, by its presence, depresses the rate of a process.

Each concept, that of limiting factors and that of retarding ones, "is but a special case of the law of toleration" (Shelford, 1913), but distinguishing between limiting and retarding factors is particularly important in the study of plant-pest control. All plant pests can be viewed as retarding factors that are alive, and people combat them by manipulations that involve very different principles. For example, pest-resistant varieties are pest-limiting factors: they decrease the amount of nourishment that is available to a population of parasites. Protective pesticidal chemicals are pest-retarding factors: they depress the rate of development of the population of pests.

The Laws of Thermodynamics

Ecosystems are characterized by the fact that energy flows through them. Solar energy and chemical energy (nutrients from the soil) are used by green plants, which provide a source of energy for the herbivores that consume them. Energy from certain herbivores (cattle, fowl) is then made available to carnivores (humans, beasts of prey). At any point in this flow (Figure 1.6), a certain amount of energy is returned as heat to the environment by the respiration of the organisms of the ecosystem. This flow of energy, its absorption and release by living things, is the central feature of the ecosystem. And the flow of energy obeys two fundamental laws: the first and second laws of thermodynamics.

The first law of thermodynamics states simply that energy can be neither created or destroyed; only its form can be altered. This is the familiar law of "conservation of energy." The amount of existing energy remains constant; physical and biological processes alter nothing more than the distribution of energy among its different forms (radiant, chemical, mechanical, atomic, and electrical).

The second law of thermodynamics specifies the direction in which energy flows during physical processes. In terms of heat, for example, the second law states that, in self-sustaining processes, heat (energy) flows from the hotter to the colder mass. Similarly, chemical energy would flow from the more-concentrated to the less-concentrated; it diffuses at each step in the flow.

It follows from the second law of thermodynamics that the energy available in ecosystems to organisms for the purpose of growth is continually diminishing. Even though none of the energy is lost, the form of the original energy is changed into unusable forms at each

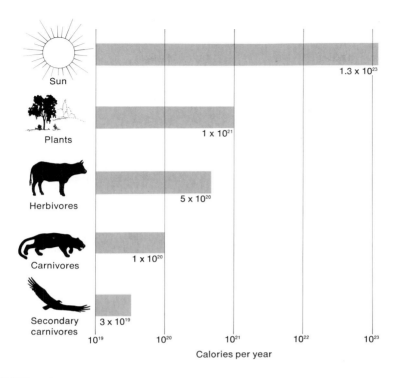

FIGURE   1.6

The use of solar energy, in sequential steps, by green plants, herbivores, and carnivores. [From "The ecosphere" by LaMont C. Cole. Copyright © 1958 by Scientific American, Inc. All rights reserved.]

point in its stepwise flow through the ecosystem. Only a small part of the Sun's energy can be trapped and used by green plants; only a fraction of the energy from green plants is in a form that herbivores can use; and very little of the energy from herbivores is available for use by carnivores (Figure 1.7). Human beings thus require a continuous input of energy to maintain themselves on planet Earth (Ehrlich and Ehrlich, 1970).

PLANT-PEST CONTROL AND ECOSYSTEMS

As indicated in the first section of this chapter, about 60 percent of the Earth's land mass is populated by green plants that people manage to some degree. By their size alone, then, agricultural and forest ecosys-

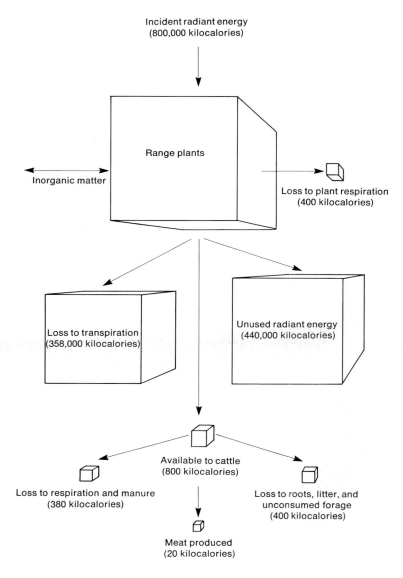

FIGURE   1.7

The continual loss of useful energy, according to the second law of thermodynamics, as it flows from the Sun through plants and cattle to human beings. Figures represent energy consumed during eight months of meat production on an acre of rangeland in California. [From "The rangelands of the western United States" by R. Merton Love. Copyright © 1970 by Scientific American, Inc. All rights reserved.]

tems play a significant role in the Earth's ecology. Whereas natural ecosystems are characterized by processes that recycle matter and energy, agricultural ecosystems all too often remove from the Earth more than they return. Energy available for agricultural use thus diminishes continually.

Plant pests, those pernicious, noxious organisms that victimize plants, also play significant roles in the ecology of the Earth. By denying us at least a third of what our plants could produce, pests accelerate the loss of matter and energy that would otherwise be useful for human purposes. The student of plant-pest control seeks to discover how losses to plant pests can be averted in the interest of maximum efficiency in crop production. But the student, knowing the laws of thermodynamics, aware of the concepts of limiting and retarding factors, and conversant with the language and science of ecology, seeks also to achieve this goal without doing lasting and irreparable harm to the environment.

LITERATURE CITED

Beirne, B. P. 1967. *Pest Management*. Cleveland: CRC Press.

Bisplinghoff, R. L., and J. L. Brooks. 1972. "Role of basic research in implementing pest control strategies." In *Pest Control: Strategies for the Future*. Washington, DC: National Academy of Sciences.

Blackman, F. F. 1905. "Optima and limiting factors." *Ann. Bot.* (London), 19:281–295.

Boughey, A. S. 1971. *Fundamental Ecology*. San Francisco: Intext.

Braidwood, R. J. 1960. "The agricultural revolution." *Scientific American*, September: 130–148.

Bunting, A. H. 1972. "Ecology of agriculture in the world of today and tomorrow." In *Pest Control: Strategies for the Future*. Washington, DC: National Academy of Sciences.

Chester, K. S. 1947. *Nature and Prevention of Plant Diseases*. Philadelphia: Blakiston.

Curtis, O. F., and D. G. Clark. 1950. *An Introduction to Plant Physiology*. New York: McGraw-Hill.

Ehrlich, P. R., and A. H. Ehrlich. 1970. *Population, Resources, Environment*. San Francisco: W. H. Freeman.

Fuller, H. J., and D. D. Ritchie. 1967. *General Botany*, 5th ed. New York: Barnes and Noble.

Janick, J. 1963. *Horticultural Science*. San Francisco: W. H. Freeman.

Janick, J., R. W. Schery, F. W. Woods, and V. W. Ruttan. 1974. *Plant Science,* 2nd ed. San Francisco: W. H. Freeman.

LeClerg, E. L. 1964. "Crop losses due to plant diseases in the United States." *Phy.opathology,* 54:1309–1313.

Metcalf, C. L., and W. P. Flint. 1932. *Fundamentals of Insect Life.* New York: McGraw-Hill.

Odum, E. P. 1971. *Fundamentals of Ecology,* 3rd ed. Philadelphia: Saunders.

Roberts, D. A., and C. W. Boothroyd. 1972. *Fundamentals of Plant Pathology.* San Francisco: W. H. Freeman.

Shelford, V. E. 1913. *Animal Communities in Temperate America.* Chicago: University of Chicago Press.

Tschirley, F. 1972. "Comments." In C. B. Browning, ed., *Systems of Pest Management and Plant Protection.* Gainesville: University of Florida, College of Agriculture.

Whetzel, H. H. 1929. "The terminology of phytopathology." *Int. Congr. Plant Sci. Proc.,* 2:1204–1215.

2

The Structure of Crop Plants

The answers to two questions mark the beginning of all wisdom; for what do we need to know of something—an object, a person, an idea, a crop plant, or its pest—except "What does it look like?" and "How does it work?" Form and function—structure (anatomy and morphology) and behavior (physiology)—are what the plant-pest-control expert must learn about the crop plants he seeks to protect and the pests he seeks to control.

Knowledge of a living organism's form, of its structure in terms of morphology itself as well as in terms of anatomy—which is the branch of morphology that treats the subject of "minute internal structure" (Fuller and Ritchie, 1967)—is the basis of taxonomy, the study of classification and identification. Anatomical and morphological features of an organism thus tell us what it looks like. Because "like begets like," organisms can be classified, according to their comparative developmental morphology, into natural or phylogenetic groups. When comparative morphological details are lacking, however, scientists must resort to artificial groupings, which arrange "look-alikes" into groups without regard to their natural relationships. Any organism, once classified into a group with its relatives, requires a name so that its identity can be established. All organisms are identified by their Latinized binomials (Chapter 1). Conventions for the naming of animals and plants are set forth in the *International Code of Zoologi-*

cal Nomenclature (Stoll et al., 1964) and in the *International Code of Botanical Nomenclature* (Stafleu et al., 1972). How these conventions are put to use is described in the next section.

CLASSIFICATION AND NOMENCLATURE

Taxonomy is the science of classification, which is the ordering of organisms into groups on the basis of their relationships, usually structural. Identification rests on a recognition of structural and functional characteristics of a taxon and is established by nomenclature, the naming of organisms. The student of plant-pest control is primarily interested in the identity (name) of crop plants and their pests, but he is also aware that the identification of these organisms derives from their classification. In Table 2.1, classifications of some plants are outlined, and several important crop plants are identified by their scientific names, or binomials (Fuller and Ritchie, 1967; Janick et al., 1974).

The authority, or person responsible for assigning a Latinized binomial to an organism, is listed immediately after the specific epithet (name of the species). The name of the authority is often abbreviated. For example, the authority "L." is the abbreviated form of Linnaeus (Carl von Linné), whose 1753 publication, *Species Plantarum,* is the starting point for the nomenclature of higher plants. The earliest published and correct name after 1753 thus becomes the valid scientific name of a plant.

As taxonomists learn more about the structure of previously named organisms, they frequently discover that the original name identified the organism erroneously and that it must be renamed according to the conventions of the *International Codes of Nomenclature.* For example, Linnaeus first named the soybean *Soja max,* but E. D. Merrill (abbreviation: Merr.) transferred that species, according to the rules of nomenclature, from the genus *Soja* to the genus *Glycine.* The soybean is thus referred to as *Glycine max* (L.) Merr., and *Soja max* L. is reduced to the rank of synonym. The code of zoological nomenclature permits, but does not require, the naming of an authority who transfers a species of animal to another genus.

Faced with the task of identifying a weed, parasite, insect, herbivore, or any other plant pest, the pest-control specialist first studies the specimens available to him, paying particular attention to morphological features. Then he turns to a taxonomic key, which may be either analytical or synoptic.

The familiar analytical key, often termed the dichotomous key, is based on single-characteristic differences at each hierarchy. For

TABLE 2.1.
A classification of plant groups.[a]

KINGDOM: **Plant** (as opposed to animal)
  SUBKINGDOM: **Thallophyta** (plants not forming embryos)
    Phyla (Divisions) 1-7: (algae)
    Phylum 8: **Schizomycophyta** (bacteria)
    Phylum 9: **Myxomycophyta** (slime molds)
    Phylum 10: **Eumycophyta** (true fungi)
  SUBKINGDOM: **Embryophyta** (plants forming embryos)
    Phylum 11: **Bryophyta** or **Atracheata** (plants lacking vascular tissue: mosses, liverworts, and hornworts)
    Phylum 12: **Tracheophyta** or **Tracheata** (plants with vascular tissues)
      Subphyla containing psilopsids, club-mosses, and horsetails
      Subphylum: **Pteropsida** (ferns and seed plants)
        Class: **Filicineae** (ferns)
        Class: **Gymnospermae** (cone-bearing plants)
          Order: Coniferales
            Family: Pinaceae
              Genus: *Pinus*
                Species: *palustris*
[Long-leaved pine is thus classified as indicated above, and is identified by its binomial: *Pinus palustris* Mill.]
        Class: **Angiospermae** (true flowering plants)
          Subclass: **Monocotyledonae** (seed with a single seed leaf)
            Order: Palmales
              Family: Palmaceae [*Cocos nucifera* L., coconut palm]
            Order: Graminales (Poales)
              Family: Graminae or Graminaceae (Poceae) [*Triticum aestivum* L., wheat; *Oryza sativa* L., rice: *Zea mays* L., corn or maize; *Saccharum officinarum* L., sugar cane]
            Order: Musales
              Family: Musaceae [*Musa sapientum* (L.) Kuntze, banana]
          Subclass: **Dicotyledonae**
            Order: Chenopodiales
              Family: Chenopodiaceae [*Beta vulgaris* L., sugar beet]
            Order: Papaverales
              Family: Cruciferae [*Brassica* spp., cabbage, cauliflower, turnip, etc.]
            Order: Rosales
              Family: Rosaceae [*Rosa* spp., roses; *Prunus* spp., almonds, apricots, cherries, peaches, and plums; *Malus sylvestris* Mill., apple]
              Family: Leguminosae [*Phaseolus vulgaris* L., common garden bean; *Glycine max* (L.) Merr., soybean; *Medicago sativa* L., alfalfa, or lucerne; *Trifolium pratense* L., red clover]
            Order: Malvales
              Family: Malvaceae [*Gossypium hirsutum* L., upland cotton]
            Order: Campanulales
              Family: Cucurbitaceae [*Cucumis melo* L., muskmelons; *Cucurbita* spp., squash and pumpkins]
              Family: Compositae [*Carthamus tinctorius* L., safflower; *Chrysanthemum morifolium* (Ramat.) Hemsl., florist's chrysanthemum]

[a]*Source:* Fuller and Ritchie (1967).

example, as indicated in the classification scheme set forth in Table 2.1, we can distinguish between the subclasses Monocotyledonae and Dicotyledonae by knowing a single morphological feature: the number of seed leaves. But it is not always so easy to distinguish between minor taxa on the basis of a single feature. A critical ("key") characteristic may be so obscure that the inexperienced person might not recognize it. Further, specimens that have been damaged might not provide information on a critical characteristic. Consequently, a synoptic key may well prove to be more satisfactory than an analytical one.

The synoptic key first lists the already-established taxa in question and identifies each taxon, usually by a number. Then the synoptic key lists a number of structural characteristics, several of which may be shared by two or more taxa, and indicates which taxa possess each characteristic. Similarities and differences between the taxa can then be discovered by comparing their shared and nonshared characteristics. A previously unidentified specimen can then be properly identified by relating several of its features to those shared by previously described and correctly named taxa. A synoptic key is compared with an analytical key in Table 2.2, in which six genera of plant-parasitic bacteria have been identified.

Aside from the advantages already stated, the synoptic key, based on many characteristics, provides more information on the structure of organisms than does an analytical key. But possibly its ultimate value is that the synoptic key is the foundation for the fascinating and expanding science of numerical taxonomy. Only by means of this new science, which relies heavily on mathematics and computer technology, can biologists even begin to handle systematically the increasing numbers of new characteristics discovered for any species or higher taxon. The theory of numerical taxonomy and application of its principles—not only to classifications of living organisms, but also to ecology, soil science, genetics, anthropology, diagnostic medicine, geology, and even political science—have been splendidly set forth by Sokal and Sneath (1963).

Knowledge of an organism's structure not only permits its classification and naming, but often suggests its function. Anatomical features, rather than gross morphological characteristics, commonly determine function. For example, xylem vessels are well adapted to water conduction in green plants; sieve tubes are adapted to the bidirectional transport of elaborated food materials; the chlorophyll-rich foliar palisade and spongy parenchyma are adapted to the manufacture of sugars by photosynthetic processes. Conversely, impairment of normal functions is often reflected by defective structural development.

TABLE 2.2.

A comparison between an analytical and a synoptic key to the genera of plant-parasitic bacteria that can be grown in culture.[a]

Analytical (Dichotomous) Key	
Bacteria facultatively anaerobic	1. *Erwinia*
Bacteria aerobic	
Endospores formed	2. *Bacillus*
No endospores formed	
Positive reaction to Gram's stain	3. *Corynebacterium*
Negative reaction to Gram's stain	
Peritrichous flagella	4. *Agrobacterium*
Polar flagella	
Yellow in culture	5. *Xanthomonas*
White in culture	6. *Pseudomonas*

Synoptic Key

Numbers[b] correspond to those identifying the genera named above.

A. Strictly aerobic	2,3,4,5,6
B. Endospores formed	2
C. Gram-positive	2,3
D. Peritrichous flagella	1,2,3,4
E. Polar flagella	5,6
F. Yellow in culture	*1*,3,5
G. White in culture	*1*,2,4,6
H. Cause soft rots	*1*,2
I. Cause vascular wilts	*1*,3,6
J. Causes tumors in plants	4
K. Cause blights and leafspots	*1*,3,5,6

[a]*Source:* R. E. Stall (personal communication).

[b]Italicized numbers mean that the characteristic is not possessed by all species.

For example, gall wasps, certain nematodes, the crown-gall bacterium, and certain fungi all induce ugly tumors in the plants they victimize. Also, the competition of weeds, feeding of vermin and insects, and parasitism by microorganisms and viruses damage the structures of the victimized plants.

The varied aspects of the form and function of such a heterogeneous mixture of living organisms as plant pests are considered in Chapters 6 through 10. Though some facets of the structure and functioning of crop plants must also receive attention in those chapters, I propose to summarize our knowledge of what crop plants look like in this chapter, and of how they work when they have not been victimized by plant pests, in Chapter 3. For how can we deal intelligently with the pests of our crop plants unless we know the innate capacities of those plants to yield a quality product in full measure?

THE STRUCTURE OF CROP PLANTS

With few exceptions, crop plants belong in two classes of the Tracheophyta (plants with vascular, or conductive, tissue): Gymnospermae (cone-bearing forest trees and woody ornamentals) and Angiospermae (true flowering plants). And crop plants in the Angiospermae belong in either of two subclasses: Monocotyledonae or Dicotyledonae. We are therefore concerned, in crop-plant production, with the structure of the so-called seed plants.

The Organs of Seed Plants

Organs are structures adapted to perform specific functions, and the major organs, or principal parts, of a seed plant are familiar to all who know even the rudiments of botany. The plant body (Figure 2.1) consists of only four major parts: roots, stems, leaves, and cones or flowers (Fuller and Ritchie, 1967). The first three are termed vegetative, the others reproductive. Eames and McDaniels (1925) have aptly described the constitution of the plant body:

> The root, stem, and leaves of a plant constitute its *organs*. These perform distinct general functions for which they are adapted by the kinds, proportion, and the arrangement of the *tissues* of which they are composed. The tissues have more restricted functions, which are determined by the kinds of *cells* which constitute them. The plant body thus consists of cells, which are aggregated to form tissues, and these, in turn, are grouped together to form organs.

Roots serve mainly to anchor the plant, to absorb water and minerals from the soil, and to store food; stems support the leaves and flowers, store food, and transport substances to and from the leaves and roots; leaves, by means of photosynthesis, manufacture food out of water absorbed from the soil and carbon dioxide from the air; cones and flowers are reproductive organs whose function is to form seeds (Fuller and Ritchie, 1967).

The Plant Body

Perhaps the most striking feature of the plant kingdom is the wide-ranging diversity of form among its members. And the several organs

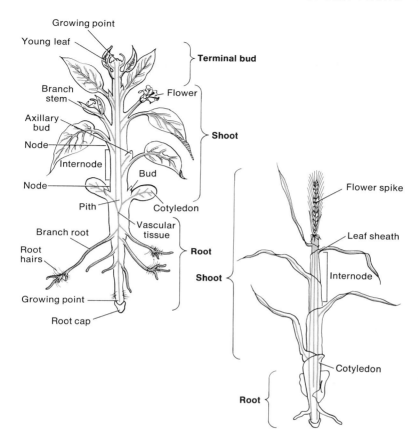

FIGURE 2.1

The principal parts of the plant body: roots, stems, leaves and flowers. Left, diagram of a dicotyledonous plant; right, diagram of a monocotyledonous plant. [From *Plant Science,* 2d ed., by Jules Janick et al. W. H. Freeman and Company. Copyright © 1974.]

of seed plants often disguise themselves almost beyond recognition. Moreover, there is very little correlation between the form of the plant body and the taxonomic position of plant species. For example: Who would suspect the familial kinship of the honey locust tree and the common bean plant? Who would notice the family resemblance between the pineapple and Spanish moss, that unique epiphyte—neither Spanish nor moss—whose graybeard pendants weirdly decorate the trees and give an eerie aspect to our southern lowlands? Who would expect, from its appearance, that the sago "palm" is not a palm at all, but a cycad? Diversity of form and complexity of structural features often mask the "simple, uniform structural plan" of the plant body, which, in essence, consists of only two components: a central axis (roots and stems) that bears lateral appendages (leaves and flowers). This reassuring and encouraging assessment is that of Eames and McDaniels (1925), on whose concise account of plant anatomy the concepts of this section on the plant body are based.

The Axis of the Plant Body

The axis, encompassing both stem and root, consists of two principal parts: a core, termed the central cylinder or stele, and a surrounding sheath, several cells thick, termed the epidermis (Figure 2.2). Conductive tissues of two kinds, xylem (primarily water-conducting) and phloem (food-conducting), make up most of the stele, whose central portion in many plants consists of soft and loose tissue termed pith. The vascular tissues are separated from the surrounding cortex by several layers of nonconductive cells termed the pericycle, which is usually delimited by a one-celled layer of tissue termed the endodermis. Vascular tissues usually occur together, with the phloem outermost. The actively dividing cells between phloem and xylem are termed the cambium; new cells that are cut off inward from the cambium develop into water-conducting xylem elements, whereas those cut off outward develop into food-conducting phloem elements. Vascular tissues nearest the cambium are termed primary, those farthest away, secondary. In section, progressing inward, the axis of the typical seed plant thus consists of epidermis, cortex, and stele; and the stele is composed of endodermis, pericycle, phloem, cambium, xylem, and pith.

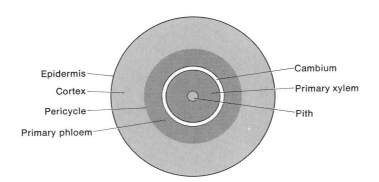

FIGURE 2.2

Locations of dicotyledonous stem tissues of the primary plant body as they appear in a transverse section.

The axis of the plant body is continuous from root to stem, and the cortex, endodermis, pericycle, and secondary phloem and xylem are directly continuous between the two organs. But primary vascular tissues, though continuous, are not directly so between root and stem. There is a transitional zone between root and stem, a zone in which the primary vascular arrangements of the root are modified to mesh with the very different arrangements of primary vascular tissues of the stem. The phloem and xylem of the root are arranged independently, in radial strands, whereas in the stem the two conductive tissues are arranged together in collateral bundles, with the phloem outermost.

The axis of a plant grows only at special growing points at its tips. These growing points consist of meristematic tissue (Figure 2.3), which is made up of actively dividing cells. The tissue at the uppermost part of the axis is termed the apical meristem, that at the lowest part, the root meristem. Other meristems of the axis include the cambium, a lateral meristem already mentioned, and the short-lived meristems situated between apical meristems, such as the intercalary meristems at the bases of leaves of the Monocotyledonae.

The growth described so far is primary growth, which increases the length of the axis, and initiates its system of branching and adding the appendages (leaves and flowers, treated in the next two sections). Although the plant body is fundamentally structured by primary growth, its bulk—especially that of its conductive tissues—is often greatly increased by a different kind of growth: secondary growth. Secondary growth does not usually add new kinds of cells and tissues; instead, after the fundamental parts derived from primary growth have matured, secondary growth increases the diameter of the organs of the primary plant body. The most striking evidence of secondary growth is found in the continuously expanding woody Gymnospermae and Dicotyledonae, but some secondary stem thickening is apparent in many herbaceous Dicotyledonae. Girth of monocotyledonous stems is not usually increased by secondary growth, but certain palm stems expand in diameter by secondary development of tissues of the pith, in which the separate bundles of conductive tissues are imbedded.

Secondary thickening of stems of the Gymnospermae and the Dicotyledonae of the Angiospermae is accomplished by the continued meristematic activity of the cambium, from which primary xylem and phloem are formed. As cells of the cambium continue to divide, secondary xylem and phloem are produced, and the primary conductive tissues of the phloem are pushed outward and crushed by the forces of growth from within. The cambium forms a continuous cylinder around the interior of the stele, not only in woody plants but also in many herbaceous dicotyledonous plants, whose primary vascular tissue is

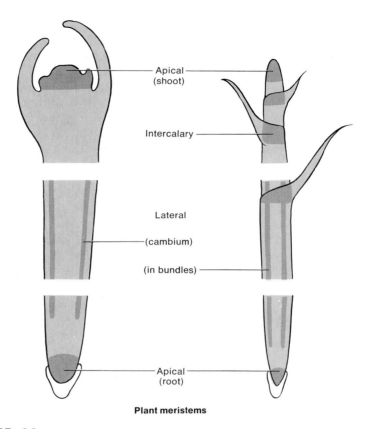

Plant meristems

FIGURE   2.3

Position of plant meristems in dicotyledonous plants (left) and in monocotyledonous plants (right).

arranged in collateral bundles around the periphery of the stele. A continuous band of cambium develops as the meristematic cambial initials between the bundles become active. This newly produced conductive tissue connects laterally with the corresponding tissues of the original bundles. A dicotyledonous plant stem is diagramed in Figure 2.2.

Secondary thickening proceeds in roots as in stems, but, because of the independence of primary xylem and primary phloem in roots, secondary thickening in roots presents an outward appearance somewhat different from that in stems. In roots, the cambium arises as strips of meristematic cells inside the groups of primary phloem. Cambial strips then extend laterally and join together to form an irregular layer, which curves outward around xylem rays and inward around the phloem. Finally, the asymmetric ring of cambium becomes cylindrical, because secondary tissue develops more rapidly in cambial segments that underlie the primary phloem than in those segments that originally develop outside the primary xylem. Figure 2.4 diagrams the root of a dicotyledonous plant.

Appendages of the Plant Body

Although a plant's principal appendages, leaves and flowers, bear little morphological resemblance to the axis (stems and roots), the appendages are, nevertheless, anatomically nothing more than lateral extensions of the axis. Leaves and flowers thus contain epidermal, cortical, and stelar tissues, structurally modified from their counterparts in the axis.

Leaves. Leaves consist of stalks (petioles), stipules, and blades, which, though usually flattened, may be needlelike, as in the Coniferales. The chief function of the leaf is to manufacture, by photosynthetic activities, complex foods from carbon dioxide, water, and minerals. Stelar tissues, consisting of vascular and other supportive tissues, form the "skeleton" of the leaf. And this "skeleton" supports the green, photosynthesizing tissues (cortex) that are enclosed by the layer of epidermal cells, which are continuous with the epidermal cells of the stem (Figure 2.5). The cortex-derived green tissue of the leaf is parenchymatous; that is, its cells are unspecialized vegetative ones with relatively thin cellulosic walls. The parenchyma between

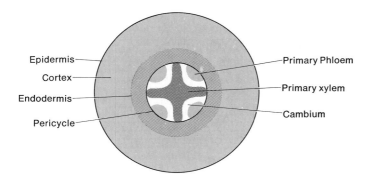

FIGURE   2.4
Locations of dicotyledonous root tissues of the primary plant body as they appear in a transverse section.

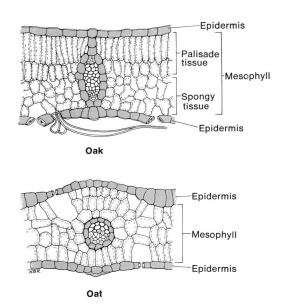

Oak

Oat

FIGURE   2.5

Drawings of cross sections through leaves of oak (top) and oat (bottom). [From *An Introduction to Plant Anatomy,* by A. J. Eames and L. H. MacDaniels. McGraw-Hill Book Company. Copyright © 1925.]

the upper and lower epidermal layers is of two types: closely packed, elongated palisade parenchyma; and loosely arranged, irregularly shaped, spongy parenchyma. Together, these two tissues make up the mesophyll.

The stelar or skeletal parts of leaves are often termed veins, which may be arranged in two ways: in parallel or in the form of a net. Parallel venation is characteristic of the Monocotyledonae, net venation of the Dictotyledonae. Since the leaf is a lateral extension of the stem, where phloem is external to xylem, it follows that the phloem is usually on the underside of the petiole and leaf, whereas the xylem is on the upper side. It is probably for this reason, according to Eames and McDaniels (1925), that the upper side of the leaf is often termed the ventral side, the lower, dorsal. The ventral surface is thus adaxial (toward the axis), the dorsal, abaxial (away from the axis).

Flowers, Fruit, and Seed.   Whereas leaves are highly modified vegetative extensions of the axis, cones and flowers (perhaps the best-disguised apendages) are modified reproductive extensions of the axis. Three kinds of flowers are diagramed in Figure 2.6. Whatever its structural sophistication, the flower is borne on a stalk (pedicel) that branches as an appendage to the axis, and the pedicel is a typical stem with a cylinder of vascular tissue or a ring of vascular bundles. The pedicel terminates in a swollen tip, the receptacle, which supports the four principal organs of the flower itself: sepals, petals, stamens, and pistil. The stelar tissue is broadened and flattened in the receptacle, and from such stelae arise the traces of vascular tissue found in the floral organs. Two of the floral organs, the stamens and the pistil, are termed essential parts of the flower. The other two organs, sepals and petals, are termed accessory parts. A flower containing all four organs is said to be a complete flower. A perfect flower contains both stamens and pistil, whereas an imperfect flower contains one or the other, but not both. A plant is termed monoecious if it bears both imperfect staminate flowers and imperfect pistillate flowers on the same plant.

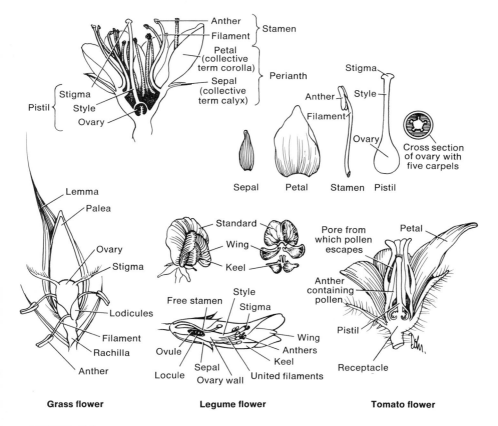

Grass flower   **Legume flower**   **Tomato flower**

FIGURE   2.6

Diagrams of flowers of grasses (left), legumes (center), and tomato (right). [From *Plant Science,* 2d ed., by Jules Janick et al. W. H. Freeman and Company. Copyright © 1974.]

Dioecious plants bear staminate flowers on one plant, pistillate ones on another.

Sepals, collectively termed the calyx, are usually green and serve to protect other floral parts. Petals, which together make up the corolla, are often brightly colored and sometimes secrete aromatic compounds and a sugary solution termed nectar. Insects, necessary for the pollination of many flowers, are attracted by the petals and their products (Fuller and Ritchie, 1967).

The stamen, usually referred to as the male reproductive part, consists of a delicate stalk (filament) at whose top is a pollen-bearing enlargement, the anther. The pistil, the so-called female part of the flower, may be either simple or compound. The pistil consists of a swollen base, the ovary, which contains the undeveloped seeds, the ovules. The ovary has above it a stalk, the style, which is enlarged at its tip into the structure termed the stigma. Flowers may be borne in clusters termed inflorescences (Figure 2.7), and the stalk of an inflorescence, which gives rise to pedicels of individual flowers, is termed a peduncle.

Following fertilization of a flower, the fruit are set, and with the maturity of the fruit, seeds are formed. A fruit, irrespective of its size and shape, is a matured ovary, whereas a seed is a matured ovule (Fuller and Ritchie, 1967). The body of a fruit develops from the ovary wall, and consists of three layers: the pericarp, mesocarp, and endocarp. A fruit may be fleshy or dry, depending on the structural characteristics of its three layers. For example, a peach consists of an outer layer of pericarp ("skin") surrounding a much thickened mesocarp ("flesh") that surrounds an extremely thick and hard endocarp ("pit"). The cavity (locule) within the endocarp contains the seed. In dry fruit, the pericarp, mesocarp, and endocarp may be indistinguishable from one another, and may appear as just a thin, dry layer around the seed-bearing locules.

A seed consists of three parts: the protective covering termed the seed coat (testa); food-storage tissue, termed the endosperm; and the immature plant itself, the embryo. Most embryos do not begin to digest and metabolize the food stored in the endosperm until they begin to germinate, but embryos of the Leguminosae (beans, peas) digest the endosperm before maturity, and the endosperm disappears before the seed leaves the plant that bore it. Food is digested from the endosperm—early or late—by the cotyledons, which, in turn, store food and are one of the three principal parts of the embryo itself. The cotyledons are the seed leaves. The other two parts of the embryo are the epicotyl, attached above the cotyledons, and the hypocotyl, attached below. The epicotyl develops into the shoot, and its growing

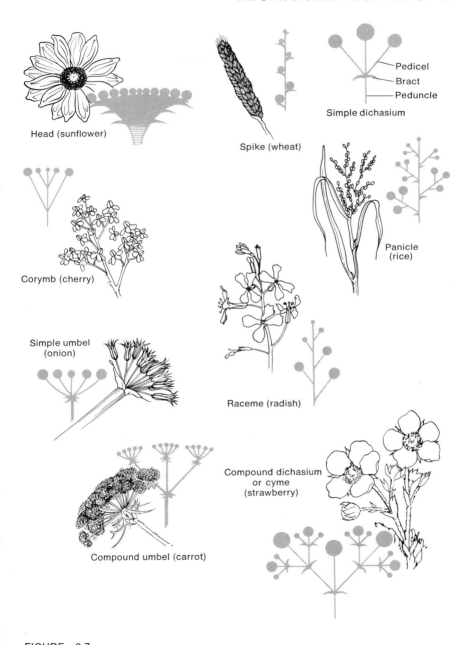

FIGURE   2.7

Diagrams of inflorescences of some familiar crop plants. [From *Plant Science,* 2d ed., by Jules Janick et al. W. H. Freeman Company. Copyright © 1974.]

tip is termed the plumule. The hypocotyl develops into the root, and its growing tip is termed the radicle.

Seeds of different plants are often very different in structure. We have already seen that seeds of legumes lack endosperm at maturity. Although most endosperms found in mature seeds are composed of parenchymatous storage tissues, endosperms of some seeds (coconut palms, castor beans) are liquid and even oily. But perhaps the most striking variation in seed structure is the number of cotyledons present. Embryos of the Gymnospermae contain three to eight cotyledons. The two subclasses of seed plants, the Monocotyledonae and the Dicotyledonae, which contain one or two cotyledons, respectively, are distinguished from each other by that single "key" characteristic.

For comparison, we may consider the seed of corn, a monocotyledonous plant, and the seed of bean, a dicotyledonous one (Figure 2.8). Being a leguminous plant, the mature bean seed lacks endosperm, but within its coat lie two much-enlarged cotyledons, which enclose the axis of the embryonic plant. The elongated epicotyl terminates in a pair of minute leaves; the short hypocotyl is inconspicuous. Beneath the coat of a corn seed lies a mealy endosperm. The embryo, which appears laterally and toward the base of the conspicuous endosperm, consists of a single shield-shaped cotyledon (termed the scutellum) and sheathed epicotyl and hypocotyl. The sheath surrounding the epicotyl (plumule) is termed the coleoptile, that encasing the hypocotyl (radicle) the coleorhiza.

THE NUCLEAR LIFE CYCLE OF SEED PLANTS

The structure of seed plants is reproduced from generation to generation by means of sexual reproductive processes and growth. Repro-

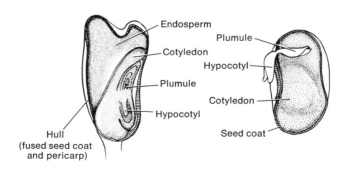

FIGURE 2.8

Drawings of a longitudinal section through seeds of corn (left) and bean (right).

duction, growth, and subsequent reproduction proceed in an ordered sequence of events termed cycles, in which there is an alternation of haploid and diploid generations. Ploidy refers to the number of chromosomes within cells, and chromosomes are the small nuclear filaments that bear the genes, the units of inheritance of particular characteristics. The characteristic chromosome number in the nuclei during the diploid (or sporophytic) generation is represented by the symbol $2n$, whereas that number in the haploid (gametophytic) generation is n. Thus, reproductive cells (gametes) bear haploid nuclei, vegetative (somatic) cells bear diploid nuclei.

As pointed out by Janick et al. (1974), "The basic difference between sexual and asexual reproduction lies in the way the chromosomes are distributed during cell division." When vegetative cells divide, the two chromosomes reproduce themselves, and daughter cells retain all the genetic material (located on the two chromosomes) characteristic of the species. Such vegetative (somatic) cell division is termed mitosis, and is exemplified by growth that takes place in meristematic regions of plants. Every cell thus bears the potentiality of every other cell of the species. In fact, Steward et al. (1964) have grown mature carrot plants, which flowered and bore seed, from a single cell that was cultured from the cambial region of the tap root.

Reproductive cells of the staminate and pistillate parts of the flowers of seed plants divide differently to produce sexual cells (gametes). Such cell division, termed meiosis, consists of two divisions and results in halving of the chromosomes. Nuclei of sexual cells are thus haploid (n). Genetic characteristics are segregated during meiosis, but are recombined later during the fertilization process.

Fundamentally, the life cycle of a seed plant proceeds as summarized by Fuller and Ritchie (1967). Pollen grains (unicellular binucleate "male" gametes) develop, following meiosis, from so-called pollen mother cells in the anthers. Meanwhile, in the pistillate flower, ovules—each containing eight haploid nuclei—form within the ovaries after meiotic cell divisions. Following pollination, which is the transfer (by wind, water, or insects, for example) of pollen to the stigma, the pollen grain germinates and sends forth a tube that grows the length of the style. The pollen tube, in which one nucleus has divided to form two sperm nuclei, enters the ovule and releases its sperm nuclei, one of which fuses with two nuclei of the ovule to form a triploid $(3n)$ endosperm nucleus. The other sperm nucleus fuses with the single egg nucleus of the ovule to form the zygote, or fertilized egg. The remaining five nuclei of the ovule and the single remaining nucleus of the pollen tube all disintegrate. By numerous mitotic cell divisions, the zygote develops into the embryo, and the endosperm nucleus develops into the endosperm of the seed.

Stripped of all structural complexities, the nuclear life cycle of any organism that reproduces sexually proceeds through a series of three events: a coming together of sexual cells (gametes) of opposite mating types, fertilization (a fusion of the haploid nuclei to form the diploid zygote), and finally meiosis (reduction), which gives rise to the sexual cells again. These events invariably occur in the sequence just indicated, and perhaps the fundamental differences between living organisms arise from variations in the time and place that each event occurs. For example, in the lowest of plants—certain fungi—meiosis is not much separated from fertilization in time, and may occur in the same organ in which nuclei of the two gametes had fused; whereas in higher plants—Angiospermae—gametes fuse in the ovules, but meiosis does not occur until after the matured seed has germinated, and the resulting plant has grown to maturity and produced its specialized organs of meiosis, the staminate and pistillate flowers. In lower forms of plant life, the gametophytic (haploid) phase of the life cycle predominates, whereas the reverse is true of the higher plants, whose axes and appendages are visible evidence of the sporophytic (diploid) phase. Intermediate forms, such as the ferns, though predominately sporophytic, produce gametophytic generations whose structure is readily detectable.

LITERATURE CITED

Eames, A. J., and L. H. MacDaniels. 1925. *An Introduction to Plant Anatomy.* New York: McGraw-Hill.

Fuller, H. J., and D. D. Ritchie. 1967. *General Botany,* 5th ed. New York: Barnes and Noble.

Janick, J., R. W. Schery, F. W. Woods, and V. W. Ruttan. 1974. *Plant Science,* 2nd ed. San Francisco: W. H. Freeman.

Sokal, R. R., and P. H. A. Sneath. 1963. *Principles of Numerical Taxonomy.* San Francisco: W. H. Freeman.

Stafleu, F. A., et al., eds. 1972. "International Code of Botanical Nomenclature." *Regnum Veg.,* 82:1–426.

Steward, F. C., M. O. Mapes, A. E. Kent, and R. D. Holsten. 1964. "Growth and development of cultured plant cells." *Science,* 143:20–27.

Stoll, N. R., et al., eds. 1964. *International Code of Zoological Nomenclature,* rev. ed. London: International Trust for Zoological Nomenclature.

3

The Functioning of Crop Plants

Knowledge of the functioning of crop plants and their pests is fundamental to our understanding of how plants are victimized by pests. Knowledge of how these organisms work, alone and in concert, is essential to the student of plant-pest control in order to assess the extent and the nature of crop-plant losses due to pests. Important aspects of crop-plant physiology and certain structural features directly related to physiology are treated in this chapter.

VITAL FUNCTIONS OF CROP PLANTS

As McNew (1950) has put it, "Chronologically, a plant may be considered as having . . . seven vital functions." These functions, correlated with the agricultural life cycle of the crop plant, include food storage (as in seeds), digestion (breakdown) and use of stored food (as in seedlings), absorption of water and minerals from the soil (by roots), meristematic activity (growth throughout the season), water conduction (in roots and stems), photosynthesis (as in leaves), and translocation of foods elaborated by photosynthesis from the manufacturing sites (leaves, for example) to storage depots (seeds, stems, and roots). These vital functions have already been used as the basis for classifying plant diseases (McNew, 1950, 1960; Roberts and Boothroyd,

1972), those examples of continuous dysfunction caused by such diverse plant pests as viroids, viruses, mycoplasmalike organisms, rickettsiae, bacteria, fungi, algae, seed plants, protozoans, nematodes, and even certain noxious insects. These vital functions might also be used to classify the different kinds of destruction brought on plants and their products by yet other pests: weeds, vermin, and injurious insects. To consider plant-pest damage in terms of these seven vital functions emphasizes the more important member of the pest-victim complex: the crop plant itself. Moreover, the farmer, on whose shoulders rests the ultimate responsibility for plant-pest control, employs his pest-control tactics chronologically, before the seed is sown, throughout the growing season, and even after the harvest has been taken and, possibly, stored against future needs. We will examine these seven vital functions of crop plants in this chapter. But an eighth vital function, respiration, deserves attention first, for this process, essential to all life, is directly or indirectly related to the seven vital functions already enumerated.

Respiration

Respiration is a chemical process of oxidation, or slow combustion (Fuller and Ritchie, 1967), of sugars to form carbon dioxide and water (Bonner and Galston, 1952); by this combustion, energy is released from the sugars, at first in the form of high-energy phosphates, and this energy is used to run the machinery of the cell (Janick et al., 1974). Respiration provides the energy for osmotic work, such as ion accumulation; for mechanical work, such as growth; and for chemical work, such as the reactions that lead to the syntheses of all carbohydrates, proteins, lipids, oils, pigments, alkaloids, latex, all compounds for which humans grow their crop plants. Respiration proceeds in all living cells.

The sum of the chemical reactions involved in a living organism is termed metabolism; so respiration and all the synthetic reactions that derive their energy from respiration are examples of metabolism. Synthetic reactions build up complex compounds from simple raw materials; such reactions are termed anabolic. Respiration, however, is a catabolic reaction, one involving the degradation of organic compounds. The general chemical equation for respiration is disarmingly simple: one molecule of glucose, in the presence of six molecules of oxygen, is oxidized to yield six molecules of carbon dioxide and six of water, plus energy:

$$C_6H_{12}O_6 + 6O_2 \rightarrow 6CO_2 + 6H_2O + energy$$

But the end products appear only after a series of stepwise anaerobic chemical reactions and a second cyclic series of aerobic (oxygen-requiring) reactions. Anaerobic respiration breaks down the six-carbon sugar into a three-carbon organic acid, pyruvic acid, with the release of some energy (Figure 3.1). In subsequent aerobic respiration, pyruvic acid is transformed into carbon dioxide and water, and additional energy is released (Figure 3.2). By the catabolism of sim-

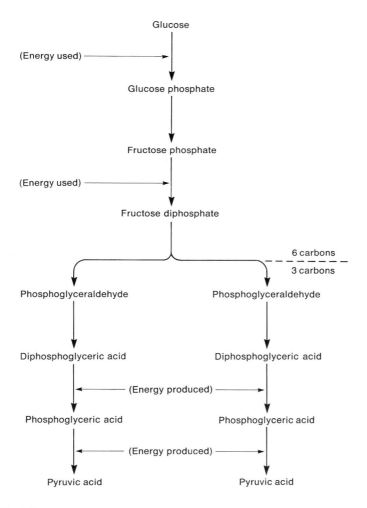

FIGURE   3.1
The pathway of anaerobic respiration (glycolysis).

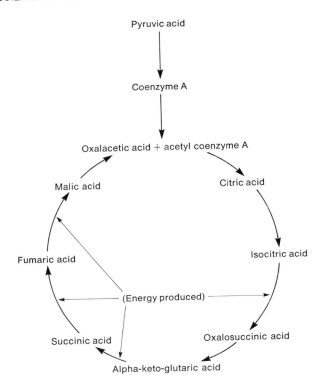

FIGURE 3.2
The pathway of aerobic respiration (oxidative respiration, the citric-acid cycle, or the Krebs cycle).

ple sugars by respiration, the physical energy of the Sun, originally trapped in green tissues by photosynthesis, is released from its storage place as chemical energy capable of performing the work of the living plant cell.

Since light is not required for respiration, the process goes on day and night. However, rates of respiration in the light may greatly exceed those in the dark. Further, not all the energy derived from the combustion of sugars is used by the plant. Instead, some of this energy is dissipated as heat. Consequently, not all respiratory reactions are beneficial, and certain crop plants sometimes fail to assimilate enough carbon dioxide by photosynthesis to compensate for that lost by respiration. Respiration in the light proceeds not only by the same reactions that occur in the dark, but also by additional respiratory reactions that presumably proceed only in the light. The term photorespiration refers to such reactions. Since photorespiration tends to affect photo-synthesis, it plays an important role in the ultimate photosynthetic efficiency of crop plants and thus affects their productivity. Generally speaking, rates of photorespiration are highest in temperate-zone crop plants, lowest in tropical grasses, but there are important exceptions to this generalization. Photorespiration, its influence on photosynthetic efficiency, and its effects on crop-improvement programs have been splendidly discussed by Zelitch (1971).

Respiration provides energy for food storage, absorption of water and minerals, meristematic activity, water conduction, and transloca-tion. Respiration itself breaks down stored food during seed germina-tion and seedling development, and it releases for use the energy originally derived from light by photosynthesis. Respiration thus plays a central role as the crop plant performs its seven other vital functions during its life cycle.

Storage of Food

Organic foodstuffs elaborated during photosynthesis, but not im-mediately respired by the plant to provide energy for anabolic reac-tions, are transported from their sites of manufacture to specialized tissues, usually parenchymatous, in which they are converted into relatively stable compounds that are stored in the receiving tissues against future needs for growth. Most foodstuffs are translocated as sugars (sucrose, for example) and stored as complex carbohydrates (starch) following polymerization in the storage tissues. The principal organs that contain storage tissues are the roots, stems, and seeds, but leaves of some crop-plant species have become adapted for food storage.

Food Storage in Roots

The root, which makes up about a third of the dry weight of the plant body, performs five functions (Fuller and Ritchie, 1967): it anchors the plant in the soil; absorbs water and dissolved minerals from the soil; conducts water and minerals to the stem above it; conducts food mate-rials elaborated in the shoot to storage tissues within the root; and, finally, works as an important food-storage organ. Roots that are much swollen with foodstuffs are termed modified or specialized roots, specialized for the purpose of accumulating and storing food.

Other familiar kinds of specialized roots are those that improve the function of anchorage, for example, the aerial roots of vines such as ivy and climbing fig, the aerial prop roots of corn, and the underground contractile roots that are associated with bulbs of such plants as onion. Some roots specialized for food storage are also specialized for asexual propagation. For example, the swollen roots of sweet potatoes and dahlia bear adventitious shoot buds, and such plants are regularly propagated by their roots.

Plant pests often interfere with food storage by roots. Fleshy roots, whether harvested for immediate human consumption or stored for later use as food or for plant propagation, can be consumed by vertebrate pests and certain insects or can decay because of plant-parasitic bacteria and fungi. Moreover, the quality and yield of the fleshy roots of such important food plants as sugar beet and sweet potato may be seriously reduced by parasitic nematodes that invade the roots while they are still developing in the field during the growing season. Of the fifteen or so most valuable food-crop plants, post-harvest pest problems with storage roots can prove serious in at least three species: sugar beet, cassava, and sweet potato. Storage roots of other important crop plants, such as carrots, turnips, and rutabagas, are equally vulnerable to attacks by a variety of pests.

Food Storage in Stems

Stems serve several purposes useful to plants: they support its leaves, flowers, and fruit; they conduct water, minerals, and elaborated foods; and they store these foods. Although food is stored in all stems, herbaceous or woody, storage is most noticeable in woody stems of such carbohydrate-rich plants as sugar cane and in specialized underground stems. Though not primarily food-storage organs, and certainly not fleshy, the woody stems of trees, which provide much of our shelter, are subject to attack by pests: insects attack living trees; termites destroy buildings; and wood-decay fungi rot lumber and buildings.

At this point, we need to digress briefly to consider several important external morphological features of plant stems: buds, nodes and internodes, spurs, offshoots, and different kinds of specialized stems. These features represent structural variations in the stem portion of the axis of the plant body, which was described in Chapter 2. Any of these variations, however bizarre it may appear, is still a stem, consisting essentially of stele, cortex, and epidermis.

Buds are "embryonic stems," consisting of a mass of meristematic tissues that "sprout" to form mature stems after the original tissues have grown by mitotic cell divisions and matured into specialized tissues for conduction and support. Buds of perennial plants function as overwintering structures, and in all plants they can initiate new growth. The location of buds is either terminal, as the growing point itself, or lateral, forming in the leaf axils (the points at which petioles join the stem that supports them). Those parts of stems from which the leaf petioles and lateral buds may develop are termed nodes, and sections of stems between the nodes are termed internodes. Buds that develop from internodes are termed adventitious; they give rise to roots—often after mechanical injury—and are important in vegetative propagation of plants by means of rooted stem cuttings.

Short branches of stems, often growing horizontally, are termed offshoots; some other common terms for them are slips, suckers, and pips. Spurs are shortened stem-branches that bear flowers and fruit. Stems of many plants, especially those of woody tissues, bear raised pores, lenticels, which are openings in the epidermis or bark, and through which gas exchange takes place. Specialized aerial stems include tendrils (climbing organs) of vines, thorns, aerial bulbs, and runners, or stolons, which produce new plants at their nodes that touch the ground (as with strawberry plants). Although the aerial stems of cacti and asparagus are well-adapted for food and water storage, stems specialized for storage are usually formed underground. Rhizomes, tubers, bulbs, and corms are examples of underground storage stems. Modified stems are depicted in Figure 3.3.

Rhizomes.   Perennial underground stems that grow horizontally are termed rhizomes, which differ from stolons, their aerial counterparts, chiefly in location. Like stolons, rhizomes produce new plants from their nodal buds, and are important propagative parts of certain plants. Slender rhizomes account for much of the propagation of turf and pasture grasses (for example, bluegrass and Bermuda grass). They also account for the rapid development of some noxious weeds, such as quack grass and Canada thistle. Fleshy rhizomes, gorged with carbohydrate, may also be important in the propagation of plants. For example, the fleshy rhizomes of the ornamental iris send forth new shoots and roots every year from the buds situated at the nodes.

Tubers.   A tuber is the greatly enlarged, fleshy tip of a rhizome. The potato is surely the most important food-crop plant whose edible portion (and propagative part as well) is a tuber. Potato tubers are subject to damage caused by disease-inducing and injury-producing plant pests from the moment they begin to develop, through the time of maturity and harvest, throughout the storage period, and, if used for propagation in the following growing season, until a new plant has developed from the propagative "seed" tuber or "seed piece."

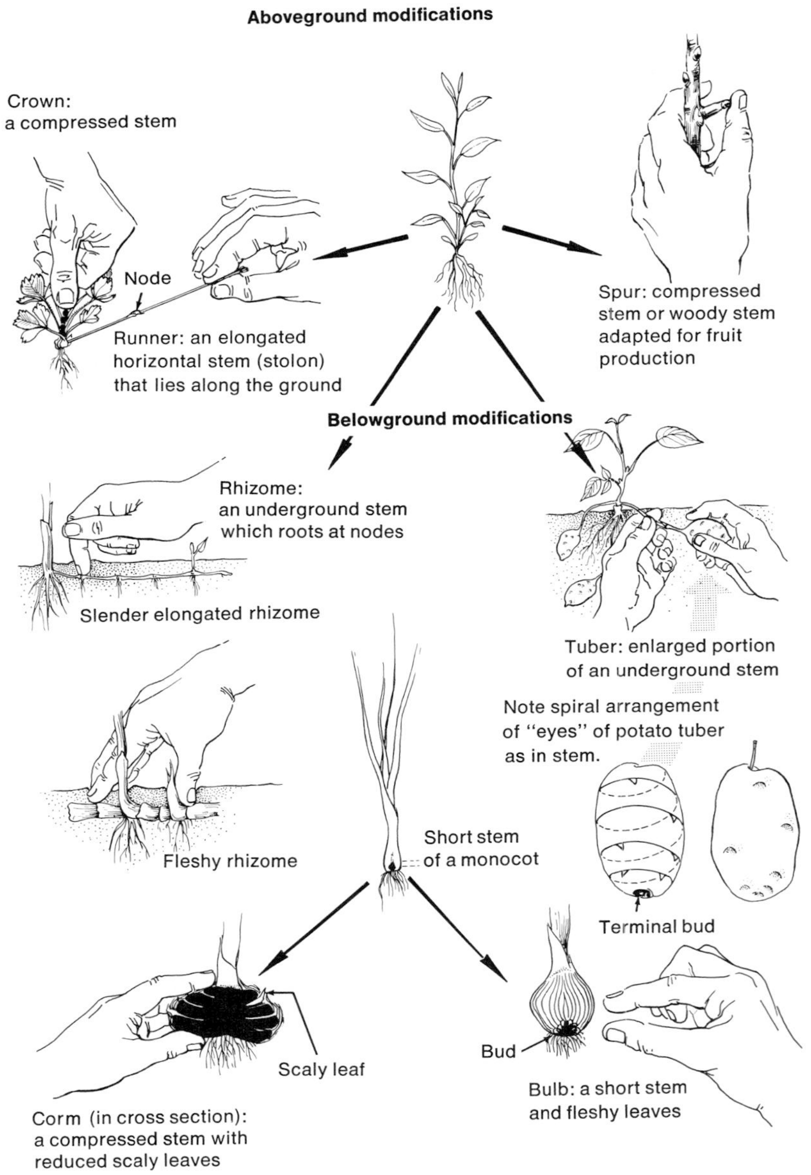

Aboveground modifications

Crown:
a compressed stem

Node

Runner: an elongated
horizontal stem (stolon)
that lies along the ground

Spur: compressed
stem or woody stem
adapted for fruit
production

Belowground modifications

Rhizome:
an underground stem
which roots at nodes

Slender elongated rhizome

Tuber: enlarged portion
of an underground stem

Note spiral arrangement
of "eyes" of potato tuber
as in stem.

Short stem
of a monocot

Fleshy rhizome

Terminal bud

Bud

Scaly leaf

Corm (in cross section):
a compressed stem with
reduced scaly leaves

Bulb: a short stem
and fleshy leaves

FIGURE   3.3

Modified stems of plants. [From *Plant Science*, 2d ed., by Jules Janick et al. W. H.
Freeman and Company. Copyright © 1974.]

Bulbs and Corms.   These globose underground modified stems are much compressed and are filled with carbohydrate-rich fleshy tissue. Unlike rhizomes and tubers, they are produced only by certain monocotyledonous plants. Most of the storage tissue of bulbs occurs in the fleshy, scale-like, overlapping leaves that enclose, at the base, a much-shortened stem with only a single bud. Important bulb crops include onions, tulips, and other ornamental plants. Corms only superficially resemble buds, for corms contain several nodes with buds, and the bulk of the storage tissue is derived from the stem. Thin, scale-like, papery leaves enclose the swollen modified stem. The ornamental plants gladiolus and crocus are propagated chiefly by corms.

Food Storage in Leaves

Although the fleshy bulb scales of such plants as onion are among the most familiar leaves that are specialized for food storage, some aerial leaves and leaf parts (petioles, for example) are sufficiently rich in stored foods to be of significance in agricultural commerce. Such widely diverse crop plants as celery, cabbage, brussel sprouts, bamboo shoots, lettuce, spinach, and rhubarb produce edible leaf parts that are grown for market in many parts of the world. And these carbohydrate-rich leaf parts are just as vulnerable to plant pests—in the field, in transit, in the markets, and in storage—as the carbohydrate-rich specialized roots and stems just discussed.

Food Storage in Flowers and Seeds

Whereas many important plant products consist of fleshy root, stem, and leaf parts, and some consist of carbohydrate-rich floral parts (cauliflower, broccoli, and artichoke), the seed is the single most important food-storage organ for the vast majority of crop plants. The seed represents the "immortality" of the plant, for the embryo, nourished by the endosperm and cotyledons and safeguarded by the surrounding seed coat, ultimately grows into a new plant, much like its parents, and thus plays the central role in the continuation of the species. From the viewpoint of human beings, who manage the plant (at least in part), the seed is often the only product for which the plant is grown. For example, rice, the cereal grains, corn, and pulses (legumes) are major human foods, and the cereal grains, corn, sorghum, and millet fatten our beef animals and nourish our herds and flocks. The food-storage function of the seed is impaired, and even the viability of the seed is jeopardized, by rats, molds, insects, and certain nematodes.

Digestion and Use of Stored Food
by Seed and Seedling

The seed cannot use the food stored within it for the germination and growth of the seedling plant until that stored food is made available by the process of digestion, which converts water-insoluble foods into usable, transportable water-soluble foods. Digestion is enzymatically catalyzed in the presence of water; hence the digestive breakdown of complex foods into simpler ones is termed hydrolysis. In the seed, hydrolysis of complex stored foods precedes respiration, which provides the energy demanded by the germinating seed (Fuller and Ritchie, 1967).

The seed uses digested foods for its germination. But a seed will not germinate until intrinsic and extrinsic factors are favorable. There are very few types of seed that will germinate as soon as they are completely formed, even when the extrinsic environmental factors (water, oxygen, temperature) are favorable. Most seed require a period of relative inactivity before their germination. This time, required by the nature of the seed itself, is termed the rest period (Curtis and Clark, 1950). Rest can be distinguished from dormancy, which also refers to a period of time during which a seed will not germinate, because, strictly speaking, dormancy is imposed on the seed by extrinsic factors; it will not germinate, for example, if kept dry. Many authors, however, do not make this distinction, but use the term dormancy to connote the period of nongerminability, no matter whether its cause is internal or external.

The field-sown seed of the crop plant performs its vital function of germination during a series of events that occur in sequence, each event depending on the preceding one. First, water is taken into the seed. This imbibed water triggers the hydrolysis (digestion) of stored food. Then the digested food materials are respired to provide energy. Finally, this energy is used for growth: the increase in numbers of meristematic cells, expansion of maturing cells, and differentiation of mature cells into tissues that are specialized in form and function. What can be seen during the germination process is the breaking of the seed coat as the seed imbibes water and as the embryo grows. Then the hypocotyl emerges and the young root can begin its functions of water and mineral absorption, which are discussed in the next section. Finally, the epicotyl emerges and grows upward, becoming the shoot, whose leaves and other green tissues begin to perform photosynthesis and whose stem begins to support the leaves, conduct water and elaborated foods, and grow in length as its apical meristem is nourished. Germinated seedlings are illustrated in Figure 3.4. De-

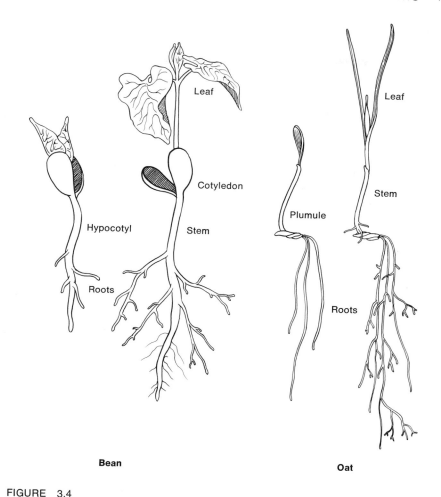

FIGURE   3.4

Seedlings of bean (left) and oat (right). [From "Seedling Development," a Turtox Class-Room Chart. Copyright © 1949 by Macmillan Science Company, Inc. All rights reserved.]

pending on the plant species, the cotyledon(s) may or may not emerge from the soil, but this feature is unrelated to the taxonomic position of the plant. For example, the cotyledons of bean emerge, those of the closely related garden pea remain underground; the single cotyledon of onion grows above the ground, whereas that of corn does not.

The young plant is termed a seedling from the time it breaks through the seed coat until it is fully dependent on food not stored, but manufactured by itself (Holman and Robbins, 1939). This delicate organism, the seedling, is highly vulnerable to destruction, not only by harsh physical factors of its environment, but also by such pests as plant pathogens, vermin, and certain insects. Moreover, the tiny seedling may be no match for its aggressively competitive, noxious relatives that we call weeds.

Absorption of Water and Minerals from the Soil

The young radicle starts its important work of anchoring the plant and absorbing water and minerals from the soil almost the moment it breaks through the coat of the germinating seed. And the root continues to perform these functions throughout the life of the plant. The function of food storage in roots has already been discussed.

All roots have three zones of development: the meristematic region, the region of elongation (enlargement), and the region of maturation. The small, densely protoplasmic cells that divide repeatedly by mitosis and that occupy the very tip of the root are responsible, ultimately, for all its growth. The new cells, destined to take on many different structural aspects and to perform various specialized functions, are all cut off by the apical initials. These working cells at the apex of the young root are thin-walled and easily damaged. If they were left unprotected, many would be destroyed in their efforts to force their way between the unmoving and abrasive particles of the soil. A fourth structural feature, the root cap, protects the vital region of meristematic cells. Protective cells that are to form the tissue of the root cap are cut off by the apical meristematic cells and are laid down in advance of the growing root. The root cap reaches full size early, and consists of a thimble-shaped mask of cells that covers the entire meristematic region of the root apex (Fuller and Ritchie, 1967). New cells are laid down at the base of the root cap at about the same rate that its outermost cells are eroded by mechanical forces as the root elongates.

Behind the meristem region, toward mature root tissue, is a region of cell elongation; this region, the cap, and the meristem account for only about two millimeters of the length of the root. After the cells have ceased to divide rapidly, they begin to enlarge, and this enlargement (predominantly along the vertical axis), like cell division, is regulated in large measure by substances termed auxins. Auxins are chemicals synthesized by plants in certain organs (usually apical meristems) that, in minute amounts, exert profound effects on growth, even in distant tissues. Analogous chemicals in humans and animals are termed hormones.

Finally, older cells differentiate into specialized root tissues, giving the root its characteristic anatomical features described in Chapter 2. Two important features of this region of maturation (Figure 3.5), root hairs and lateral roots (branch roots), deserve attention at this point. Both occur in the younger parts of the region of maturation. Root hairs are slender protuberances, a few millimeters long, of individual epidermal cells. They develop profusely from the entire circumference in a zone several millimeters long. Hairs are the chief absorbing structures of roots, increasing the absorptive surface of a root by as much as 20-fold (Fuller and Ritchie, 1967). Water is absorbed by osmosis, the soluble minerals by diffusion.

Roots develop into their familiar, far-reaching systems by branching, the production of lateral roots. The system is termed a taproot when the original seedling root develops into the main root. A fibrous root system results when the original root ceases growth in the very young plant and the system is formed by branch and adventitious roots originating from the primary root or, as in many grasses, from stem parts. Many woody plants possess a taproot system that is augmented by a shallow fibrous system of feeder roots.

Each branch root arises in the pericyclic tissue of a mature root. Groups of cells become meristematic and produce the same regions in the branch root as those just described for the primary root. The branch root elongates, pushing aside and crushing cortical cells, until it finally breaks through the ruptured epidermis. Branch roots emerge from the region of maturation just behind the root-hair zone.

It is readily apparent how plant pests damage root systems of crop plants and impair their work of absorbing water and dissolved minerals: vertebrate pests may eat the roots of plants; insects may bore in them and destroy vital tissues; and weeds may overrun the soil region intended for the crop plant. Plant-parasitic nematodes and other microscopic and submicroscopic plant pathogens (fungi, bacteria, and viruses, for example) may parasitize the living cells of roots. Plant-infecting fungi, and perhaps bacteria, lurking in the soil can enter roots through the tiny ruptures created in the epidermis when branch roots break out. Plant-parasitic nematodes use their feeding organs (stylets) to pierce the tender epidermal cells in juvenile roots. Also, some bacterial and fungal pests are able to digest the thin walls of root hairs and then grow into the epidermal cells and parasitize them. The root system of a plant is thus vulnerable to attack by a wide array of noxious organisms, and protection of the root system from the damage inflicted by pests is always difficult and expensive, sometimes impossible.

Not all root-invading microorganisms harm the plants whose roots harbor them. For example, bacteria of the genus *Rhizobium* infect

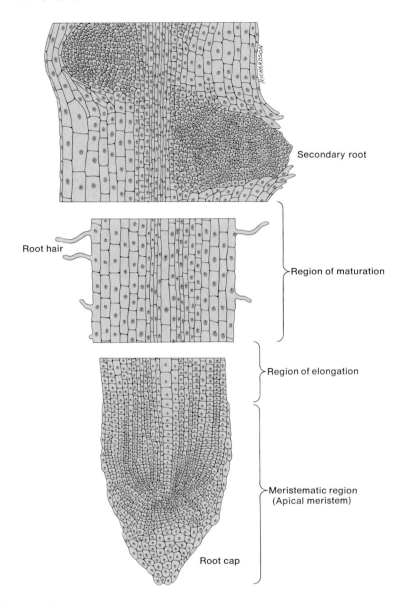

FIGURE   3.5

Longitudinal section through a plant root. [From "Root Anatomy," a Turtox Class-Room Chart. Copyright © 1937 by Macmillan Science Company, Inc. All rights reserved.]

roots, particularly those of the Leguminosae, and induce lateral galls composed of spongy host tissue and masses of bacterial cells. Significantly, these bacteria "fix" nitrogen from the air, convert it to forms the plants can use, and provide that usable nitrogen to the plant. Another example of beneficial infections: certain fungi (many of the mushroom type) invade plant roots benignly and interact with cortical tissues to increase their absorptive capacity; such infected roots are termed mycorrhizae ("fungal roots"), and occur not only among woody plants, but also among many herbaceous ones. Although nitrogen-fixing bacteria and mycorrhizal fungi are infective, since they colonize the uses of a living host, they are not pathogenic; they do not cause disease (continuous dysfunction), but actually help their host plants perform their functions. Both members of the association are benefited; each is nourished by the other, which, in turn, is assisted in its own work. The term mutualism is used to designate symbiotic food relationships mutually beneficial to both members of interacting populations.

Meristematic Activity, Growth, and Differentiation

Unlike animals, seed plants are characterized by unlimited growth, which results from the mitotic cell divisions in the meristematic tissues. Cells grow by enlargement once they cease to divide and are, therefore, no longer termed meristematic. Finally, cells mature and differentiate into cells specialized to perform the separate functions of the plant. Growth by cell enlargement is under the control of auxins, as is growth by cell division, though the mechanisms by which auxins control growth are poorly understood. The differentiation of cells into specialized tissues is even less well-understood. What makes some cells develop into cortical tissues, nearby ones into xylem, phloem, and pericycle of the stele? No one really knows, but F. C. Steward and his colleagues at Cornell University have shown that any meristematic cell has the innate capacity to give rise, upon subsequent cell divisions, to every different kind of tissue known. This capacity, termed totipotency (Steward, 1963; Steward et al., 1964), was demonstrated by means of an elegant experiment in which a plant of carrot was grown to maturity from a single cultured cell derived originally from the cambium of the swollen tap root.

Meristematic activity is adversely affected, directly and indirectly, by almost all kinds of plant pests. For example, herbivorous animals consume buds and other meristematic tissues; certain insects, fungi, and bacteria attack and kill buds; and the feeding of plant-parasitic nematodes in or near meristematic tissues may cause cell divisions to cease. Also, some pests—notably bacteria, fungi, nematodes, and insects—greatly stimulate growth by cell division and cell enlargement, inducing the formation of disruptive galls (tumors) in their crop-plant victims. Gall formation results when the pest (in some way that is not well-understood) stimulates the mature cells of the victim into reverting to a meristematic state; the pest need not occupy the meristems themselves for the stimulation to happen. In fact, colonization of meristematic plant tissue by parasitic, gall-inducing, or necrogenic pests may not lead to serious disruption of growth. For example, certain smut fungi that parasitize cereal grains infect embryos and apical shoot meristems in a benign manner; they do not inflict great damage until they invade and destroy the enlarging tissues of the immature seed. Also, most plant viruses do not invade the apical initials of meristems, but those that do (for example, tobacco ringspot virus) seem incapable of causing severe symptoms after they have invaded those cells (Roberts et al., 1970). I suspect that the meristematically active apical initials have an innate tolerance (or an innate potential for developing tolerance) to parasitic attacks by viruses or other parasitic disease-causing agents. The nature of such preexisting or pathogen-induced tolerance in meristems, if either exists, is certainly not known.

Water Conduction and Transpiration

Plants evaporate (transpire) stupendous volumes of water during their life cycles. For example, a single corn plant transpires at least 50 gallons of water during its four-month growing season. This means that the equivalent of at least 11 inches of rain is required to supply the water lost annually by transpiration in a field of corn (Curtis and Clark, 1950). Water, coming only from the soil, is essential to all plant metabolism, and it would seem that the loss of water from the leaves could be nothing but harmful. But transpiration has a modest (up to 5°C) cooling effect on leaves, whose temperatures might otherwise exceed those tolerated by the enzymes that catalyze all chemical reactions in them. Also, transpiration may indirectly influence absorption and transport of minerals, although transpiration is not necessary for either process. All in all, the process of transpiration, the loss of water in vapor form, may be of little or no benefit, and may even be more harmful than helpful (Curtis and Clark, 1950).

Whatever its advantages or disadvantages to the plant, water loss is a fact of life in the green plant. Water lost by transpiration must be replenished from the supply in the soil; otherwise the plant wilts and dies. The supply of life-giving water is distributed throughout the plant by its network of xylem. Water-deficits in leaf cells cause water to move osmotically from the ends of the xylem of leaf veins, and a strong "pull" is exerted on the water in those veins. Because of the enormous cohesiveness of water, this "pull" is transmitted throughout the xylem, even to young xylem elements in the growing roots. Water seems, therefore, to be pulled upward from the soil through the roots, stems, and leaves (Bonner and Galston, 1952).

Although a variety of plant pests may interfere mechanically or indirectly with water-conduction in crop plants, the bacteria and fungi that cause vascular-wilt diseases are the only ones that attack the xylem tissues directly, to the exclusion of other plant tissue systems. The agents of vascular-wilt diseases parasitize living cells of the xylem parenchyma and occupy the open spaces of the specialized tracheids and vessels of the xylem. The water flow is impeded, not only by the bacteria and fungi, but also by the accumulation of host-plant products, which result from partial digestion of cell walls in the water-conducting elements of the xylem.

Photosynthesis

All life on planet Earth ultimately owes its existence to photosynthesis, the reaction unique to green plants, which converts the physical energy of light into the chemical energy of carbohydrates. Light energy is absorbed by chlorophyll, the green plant pigment, which forms in discoid structures (chloroplasts) in plant cells, particularly in the palisade and spongy parenchyma of the leaves. The overall reaction for photosynthesis,

$$6CO_2 + 6\ H_2O + energy \rightarrow C_6H_{12}O_6\ (sugar) + 6O_2$$

would appear to be respiration in reverse, and, in the broadest sense, it is, for photosynthesis traps energy, whereas respiration releases energy used in the metabolic work of living cells. But, like respiration, photosynthesis is far more complicated than its overall reaction indicates.

Photosynthesis proceeds in two phases, the light and the dark. Light is trapped and water is cleaved into its component parts, hydrogen and oxygen, in the light phase. In the dark phase, the hydrogen is transferred to organic compounds and, finally, sugar is formed by a complex series of reactions. The details of photosynthetic processes, far beyond the scope of this book, are set forth in articles (Levine, 1969), and books (Rabinowitch and Govindjee, 1969; Leopold and Kriedemann, 1975).

The raw materials for photosynthesis are the water within the leaf and the carbon dioxide from the surrounding air. The rate of photosynthesis may be limited by many factors: temperature, chlorophyll content of the leaves, water, light intensity, and carbon-dioxide concentration. Even given adequate light and appropriate temperature, the normal green leaf attached to a healthy plant could not obtain sufficient carbon dioxide to manufacture sugars if it did not have stomata, a morphological feature of the leaf epidermis not yet mentioned. A stoma is an opening between two specialized epidermal cells termed guard cells, and most plant leaves have more than 100 stomata per square millimeter of lower leaf surface (Fuller and Ritchie, 1967). Fewer stomata occur on upper leaf surfaces, where the waxy cuticle over the outer cell walls of the epidermis is much thickened. Water vapor is transpired, and the oxygen required for respiration and the carbon dioxide required for photosynthesis are taken in through open stomata.

The mechanisms that make stomata open in the light and close in the dark are not understood, but, because of guard-cell structure, stomates will open when guard cells are gorged with water and close when water has been lost from the guard cells. Bonner and Galston (1952) have described stomatal movements:

(a) That part of the guard cell which abuts on the stomatal pore is thicker than the remainder of the wall. (b) Thus, when the cell becomes turgid, the thin peripheral wall is caused to bulge outward, distorting and pulling the thick elastic inner wall along with it. This opens the stomatal pore. (c) When the cell becomes flaccid, the thick, inner elastic wall contracts, thus closing the stomatal pore.

Photosynthesis is disrupted in various ways by plant pests. Weeds may shade crop plants; herbivores consume photosynthesizing tissues; insects chew and pierce leaves; and plant-pathogenic bacteria, fungi, and even certain foliar nematodes kill leaf tissues, causing the blights, leafspots, mildews, and rusts. Plant viruses and mycoplasmalike organisms also reduce photosynthetic activities, but they do so because they live and are transported in the phloem tissues of their hosts, and exert harmful effects on the vital function of translocation of elaborated food materials.

Translocation of Elaborated Food Materials

The sugars manufactured by photosynthesis are not available for respiration to provide energy for growth unless they are transported to the meristems, where growth proceeds. Moreover, sugars are not available for conversion to starch in seed and other storage organs unless they are transported to them. The phloem is the tissue of transport, and it consists of four kinds of cells: sieve tubes (the long-distance conductive cells), small companion cells (bordering the sieve tubes), fiber cells (for support), and the storage cells termed phloem parenchyma (Fuller and Ritchie, 1967). The cylindrical sieve tubes are much elongated along their vertical axes, and are connected to one another by mitered endwalls (sieve plates) that contain large perforations. The cytoplasm of the sieve tubes is continuous, appearing as thin strands (termed plasmodesmata) extending from cell to cell through the pores of the sieve plates. Plasmodesmata also connect the cytoplasm between adjacent cells in all mature plant tissues that are composed of living cells.

The manner in which materials are moved in the phloem remains controversial, if not downright mysterious. Although there is a net flow downward from leaves to roots, food materials move bidirectionally in the phloem. In trees, for example, nitrogen moves upward, not only in the water stream of the xylem (as it does in herbaceous plants), but also in the phloem. At the same time, sugars are moving downward in the phloem of trees. Bonner and Galston (1952) question: "How can the upward-moving nitrogen buck the downward pressure flow in the phloem?"

Though it cannot account for bidirectional flow in the phloem, the pressure-flow hypothesis seems a favorite of many plant physiologists. By this hypothesis, a mass flow of solutes takes place in the sieve tubes, and the force that accounts for the flow is the difference in turgor pressure between manufacturing cells (high turgor pressure) and the receiving cells (low turgor pressure). But the pressure-flow hypothesis cannot account for the simultaneous bidirectional flow of materials, and the evidence for such two-way flow—even within individual sieve tubes—continues to mount (Canny, 1971). This strengthens the notion that phloem transport may proceed not by mass flow of solutes but, most likely, by the streaming of protoplasm within and between living cells. The protoplasmic-streaming hypothesis states that "solutes are carried from one end of the cell to the other and probably even through the pores of sieve plates in the fluid strands or layers, and the movement itself is brought about by living protoplasm" (Curtis and Clark, 1950). This hypothesis would account for: (1) simultaneous bidirectional flow, (2) independent translocation of different particles (Roberts, 1952), and (3) the fact that velocity and distance are independent of osmotic pressures (Curtis and Clark, 1950). Its major weakness may lie in its inability, according to some plant physiologists, to account for the volumes and velocities at which materials flow in the phloem.

No matter what the mechanism of phloem-transport is, the fact remains that translocation is a vital function, one essential to the productivity of crop plants. Pests that interfere more with translocation than with other vital functions are few in number but often devastating in their effects. Perhaps the most important plant pests that impair the functioning of the phloem are the parasitic seed plants of the genus *Cuscuta* (dodders), the mycoplasmalike organisms that induce the "yellows" diseases, and certain plant viruses. All plant viruses that move systemically through their hosts do so in the phloem. Moreover, some plant viruses, like the mycoplasmalike organisms, are restricted to phloem, where they replicate themselves at the expense of the living protoplasm of the sieve tubes, companion cells, and phloem parenchyma.

COMMENT

The accounts of structures and nuclear life cycles in Chapter 2 and the account of vital functions of seed plants in this chapter provide some concept of what we might expect from our crop plants. We know something of what they should look like and of how they should work, and such knowledge of form and function is needed to understand the nature and extent of crop losses due to pests. From this point on, the book will be concerned principally with the plant pests themselves, their populations, how they can be controlled, and the roles they play in world crop production.

LITERATURE CITED

Bonner, J., and A. W. Galston. 1952. *Principles of Plant Physiology.* San Francisco: W. H. Freeman.

Canny, M. J. 1971. "Translocation: mechanisms and kinetics." *Ann. Rev. Plant Physiol.,* 22:237–260.

Curtis, O. F., and D. G. Clark. 1950. *An Introduction to Plant Physiology.* New York: McGraw-Hill.

Fuller, H. J., and D. D. Ritchie. 1967. *General Botany,* 5th ed. New York: Barnes and Noble.

Holman, R. M., and W. W. Robbins. 1939. *A Textbook of General Botany.* New York: Wiley.

Janick, J., R. W. Schery, F. W. Woods, and V. W. Ruttan. 1974. *Plant Science,* 2nd ed. San Francisco: W. H. Freeman.

Leopold, A. C., and P. E. Kriedemann. 1975. *Plant Growth and Development.* New York: McGraw-Hill.

Levine, R. P. 1969. "The mechanism of photosynthesis." *Scientific American,* December, 58–70.

McNew, G. L. 1950. "Outline of a new approach in teaching plant pathology." *Plant Dis. Rep.,* 34:106–110.

————. 1960. "The nature, origin, and evolution of parasitism." In J. G. Horsfall and A. E. Dimond, eds., *Plant Pathology: An Advanced Treatise,* vol. 2, 19–69. New York: Academic Press.

Rabinowitch, E., and Govindjee. 1969. *Photosynthesis.* New York: Wiley.

Roberts, D. A. 1952. "Independent translocation of sap-transmissible viruses." *Phytopathology,* 42:381–387.

————, and C. W. Boothroyd. 1972. *Fundamentals of Plant Pathology.* San Francisco: W. H. Freeman.

————, R. G. Christie, and M. C. Archer, Jr. 1970. "Infection of apical initials in tobacco shoot meristems by tobacco ringspot virus." *Virology,* 42:217–220.

Steward, F. C. 1963. "The control of growth in plant cells." *Scientific American,* October, 104–113.

————, M. O. Mapes, A. E. Kent, and R. D. Holsten. 1964. "Growth and development of cultured plant cells." *Science,* 143:20–27.

Zelitch, I. 1971. *Photosynthesis, Photorespiration, and Plant Productivity.* New York: Academic Press.

4

Dynamics of Plant-Pest Populations

Having enumerated plant pests and examined their place in agricultural and natural ecosystems, and having reviewed the anatomy and physiology of crop plants, we now turn to the topic of plant pests in terms of populations of noxious organisms. Prerequisite to an understanding of the elements of plant-pest control (Chapter 5), I believe, is some knowledge of populations and their dynamics, the ways in which they change in size (Ehrlich and Ehrlich, 1970).

Dynamics, in the general sense, refers to the moving forces in any field, to the laws that govern those forces, and to the ways such forces shift or change in relation to one another. Dynamics of populations, then, is concerned with the forces that operate to affect the numbers of individuals of a related group of living organisms. The dynamic features of a population determine its growth and regulation, affect its periodicity and stabilization, and govern its reproduction and territoriality (Boughey, 1973).

The far-ranging subject matter of population dynamics makes it difficult to give a concise definition of the term. Some would restrict its meaning. For example, Seinhorst (1970) limits population dynamics to studies of "the operation of the characters of organisms that determine population changes and their interaction." He would exclude, from the field of population dynamics, studies of external factors (even the host itself) that affect the population of nematodes. Boughey (1971) has proposed a widely accepted definition of population dynamics, which is "the study of the growth and structure of populations, together with the factors that regulate their size and cause fluctuations in their density." An even more concise definition has been proposed by Zadoks (1970), who paraphrases the definition of demography (Boughey, 1971) and defines population dynamics as the study of fluctuations in the numbers of individuals within populations. This brief definition properly emphasizes the quantitative foundation of the topic and correctly states that we are concerned with changes in numbers over periods of time, but with changes whose fluctuations are not necessarily rhythmical.

Although Zadoks' definition does not refer to the meaning of the word dynamics, his is the basis for the one used here. Let us consider population dynamics to be the study of the moving forces behind fluctuations in the numbers of living individuals within populations. The body of this chapter treats several important aspects of population dynamics that relate to the subject of plant pests and their control.

THE RISE AND FALL OF POPULATIONS

Populations of all living organisms, including plant pests and their crop-plant victims, rise and fall with time. The numbers of a species increase as long as intrinsic and extrinsic factors favor that species. But as one or more factors become limiting or retarding to the growth of a population, the rate of that growth is decelerated, often to the point that the growth rate is negative, whereupon the population declines. Concepts of population dynamics are treated in this section.

Basic Concepts

It is axiomatic that a population of individuals consists only of the living. Population dynamics is not concerned, at least at any given time, with the unborn or the dead. But the size of a population varies, during an extended period, with birth rates (natality) and death rates (mortality). Populations grow in numbers according to natality and decline according to mortality. The interrelationships between these two factors, therefore, ultimately determine the population dynamics for a stationary or sedentary species. But a third factor, migration, often governs the dynamics of a population. For example, emigration from Ireland and immigration into the United States, following the blight-induced potato famine of the mid-nineteenth century, decreased the population of Ireland and increased that of the United

States. Birth rates, death rates, and even migration determine the population dynamics of species of plant pests. Strictly speaking, migration connotes, among animals, a periodic moving from one region or climate to another for the purpose of feeding or breeding. Though some vermin migrate in the literal sense, most plant pests do not migrate, but are transported by another agent (wind, ocean currents, humans) from one region to another. Populations of plant pests in particular regions, then, may fluctuate according to transportation or migration as well as according to their rates of birth and death.

The Growth of Populations

Growth in populations is simply the increase in numbers of individuals, and it is the anticipated growth of populations of plant pests that strikes fear into the hearts of the agriculturists and moves them to act in defense of their crop plants. If they are to control a plant pest, farmers must either reduce its initial population to a safe level or must slow down the rate of growth in the pest population, to prevent it from building up to damaging proportions before the harvest is taken and safely sold, shipped, or stored.

Biotic Potential

If a population could continue to grow indefinitely and unchecked, it would, in theory, reach its full potential size. "The continuing reproduction of the organism at a maximum rate under favorable conditions" (Boughey, 1971) is termed the biotic potential of that organism. The intrinsic rate of increase (r) in a population can be expressed by the equation:

$$rN = dN/dt$$

or, integrated,

$$N_t = N_0 e^{rt}$$

where N_t is the number of individuals at a specified time, N_0 is that number at time zero (0), e is the base of natural (Naperian) logarithms (2.718), and t is elapsed time. By this equation, a plot of population density (N) against time (t) yields a J-shaped curve, which denotes an exponential rate of growth and depicts biotic potential (Figure 4.1). In nature, full biotic potential is never achieved, for conditions favorable to the reproduction of an organism never remain favorable indefinitely. Instead, factors unfavorable to maximum reproductivity come

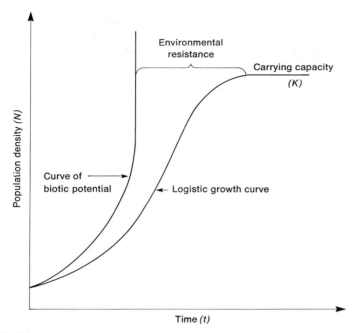

FIGURE   4.1

Theoretical curves for the growth of populations. The difference between biotic potential and carrying capacity is expressed in terms of environmental-resistance factors. [From *Ecology of Populations* by A. S. Boughey. The Macmillan Company. Copyright © 1973.]

into play, even at the outset, and limit or retard the rates of growth in populations. A given environment thus can support only a certain number of members of a given population and usually at a reduced rate of growth.

Carrying Capacity

A population, on reaching such a size that the limiting and retarding factors of its environment prevent its further increase, is said to have reached the carrying capacity of its environment. The limiting and retarding factors are referred to as environmental-resistance factors, and together can be designated by the letter K. Mathematically, the J-shaped theoretical curve for biotic potential can be corrected for environmental-resistance factors by the insertion of K into the original equation:

$$dN/dt = rN^{(K-N)/K}$$

A plot of population density against time now yields a logistic or sigmoid growth curve, whose asymptote gives the carrying capacity. Environmental resistance is depicted by the difference in slope between the steeper exponential curve and the flatter logistic one (Boughey, 1973).

The Decline of Populations

Not all populations of living things maintain a constant density once they have reached the carrying capacity of their environments. When death rates exceed birth rates, the population will decline. A natural population declines when it faces exhaustion of essential food sources, accumulation of toxic products (such as the alcohol produced by yeasts as wines are made), overcrowding of living space, or a burgeoning of natural enemies.

The rise and fall of populations of living organisms is profoundly related to the capacity of individuals to survive. For different organisms, death rates and birth rates vary widely with time during the life of an individual member of a population. For example, most starved fruit flies survive for about the same time. In countries where the rates of infant mortality are low, most men and women enjoy a life expectancy very much as indicated by actuarial tables. But individuals of many populations of green plants and of invertebrate animals die in "infancy" or while in a juvenile stage of development. Survivorship patterns for such organisms are, therefore, very different from those for human beings. Survivorship of hydra is intermediate between the two extremes just cited; the mortality rate for hydra is almost constant throughout the life of the animal. Deevey (1950) has illustrated the various kinds of mortality rates by means of survivorship curves (Figure 4.2).

INTERACTIONS BETWEEN POPULATIONS

Dynamic populations of living things fluctuate in response to environmental-resistance factors, and these limiting and retarding factors include all things, animate and inanimate, that make up the environment of the population in question and that might exert an influence on that population. Edaphic factors (soil type, soil reaction, soil temperature, soil moisture, soil fertility, etc.) and meteorological factors (temperature, rainfall, relative humidity, light quality and intensity, etc.) are among the most important environmental-resistance factors (K) that affect plant pests, victimized plants, and their interac-

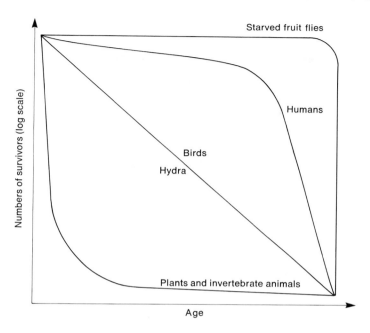

FIGURE   4.2

Survivorship patterns of different populations. [From "The probability of death" by Edward S. Deevey, Jr. Copyright © 1950 by Scientific American, Inc. All rights reserved.]

tions. Another environmental-resistance factor, interference between different populations, is of paramount importance for plant-pest control. It cannot be otherwise, for the plant-pest-control specialist always deals with two populations—that of the pest and that of its victim—and must deal with environmental-resistance factors that influence each population separately, as well as with those factors that influence the interactions between the two populations.

Interactions between two populations of living organisms can be arranged in a graded series in terms of how dependent one population is on the other. At one extreme, interactions may be unfavorable to both populations; at the other, they may be favorable to both. Between these extremes, we find some interactions favorable to one, unfavorable to the other. All such interactions have been classified by type, and scientific terms have been coined to designate each interaction (Odum, 1971). In competition, both populations, designated *A* and *B*, are affected unfavorably, and the one most severely affected may be driven, by the other, from its ecological niche. At the other extreme (mutualism or symbiosis), the interaction is not only favorable to both populations, but obligatory to both. That is, neither A nor B can exist without the other. Between competition and mutualism lie five intermediate kinds of interactions: protocooperation, commensalism, amensalism, parasitism, and predation. In protocooperation, both A and B are favored, but the interactions between A and B are not obligatory to either party. Interactions between commensals are favorable to one population, that of the commensal, but the other population, that of the host, remains unaffected. In amensalism, the interaction is inhibitory to one population, whereas the other member of the pair is favorably affected. Finally, in parasitism and predation, the interactions are favorable to parasite and predator, unfavorable to host and prey. These interactions are set forth in Table 4.1.

The student of plant-pest control must deal with species-population interactions of three categories just described: competition, parasitism, and, in a manner of speaking, predation. Of all the plant pests, only weeds qualify as competitors of crop plants. What may be

TABLE   4.1.

Relations between two interacting populations.[a]

Nature of interaction	Effect[b] of interaction on		General result of interaction
	Population A	Population B	
Neutralism (A and B independent)	0	0	Neither population affects the other
Mutualism (symbiosis) (A and B partners)	+	+	Obligatory for both
Protocooperation (A and B cooperators)	+	+	Facultative for both
Commensalism (A, commensal; B, host)	+	0	Obligatory for A
Amensalism (A, amensal; B, inhibitor)	−	0	A inhibited
Parasitism (A, parasite; B, host)	+	−	Obligatory for A
Predation (A, predator; B, prey)	+	−	Obligatory for A
Competition (A and B competitors)	−	−	Most-affected population is eliminated

[a]*Sources:* Modified after Odum (1971) and Boughey (1971, 1973).
[b]Read the symbols as follows: +, favorable; −, unfavorable; 0, unaffected.

overlooked is that crop plants are also competitors of weeds, for, in competition, both interacting populations—weeds, as well as crop plants—are adversely affected (Table 4.1). The job of the pest-control specialist is to give such a competitive advantage to the crop-plant species that the weed species are unfavorably affected and are, for all practical purposes, eliminated from the ecological niche in question, the farmer's field.

A wide array of noxious organisms qualify as parasites. In sum, they include all causative agents (pathogens) of infectious disease (continuous dysfunction caused by transmissible agents). Thus, among plant parasites are various viroids, viruses, mycoplasmalike organisms, rickettsiae, bacteria, actinomycetes, fungi, a few algae, protozoans, nematodes, and even certain kinds of insects. Whatever the parasite, its population is invariably affected favorably, whereas that of its victimized host is affected unfavorably. The specialist in plant-disease control assumes the awesome responsibility of so checking the onslaught of parasites that the intended host victim is little affected and its actual productivity approaches anticipated and hoped-for yields.

In the broadest sense, the interaction between populations of a parasite and a host may be considered obligatory for the parasite (Odum, 1971), but, strictly speaking, such an interaction is not always obligatory. There are degrees of parasitism, and not all parasites are obligate ones, that is, totally dependent on their hosts for their sustenance. Some, such as the wilt-inducing fungus, *Fusarium oxysporum*, are parasitic in living host plants but can also grow saprophytically, nourishing themselves on the dead debris of crop-plant and weed vegetation. Such organisms are termed facultative saprophytes. Other parasites (the seedling-blight fungi *Pythium debaryanum* and *Thanatephorus cucumeris*, for example) are really saprophytic organisms that qualify as parasites only after penetrating the unprotected juvenile stems of seedling plants. Such organisms are termed facultative parasites, and their populations may be maintained at high numbers in farm and forest soils for extended periods even in the absence of suitable crop-plant hosts. The practitioner of plant-pest control, to succeed, must cope with parasites, obligate and otherwise, and this task is often made more difficult by the fact that some parasites can live and grow in numbers even in the absence of their host plants.

Two final species-population interactions with which the plant-pest-control specialist must contend are those between crop plants and herbivores and between crop plants and harmful insects. As in predation (Table 4.1), these interactions are favorable to one population, the insect or the vertebrate pest, and unfavorable—even disastrous—to the other, the crop-plant population. No satisfactory terminology has been proposed to identify interactions between crop plants and either insects or vermin. Predation, which these interactions resemble, designates reactions between predators (beasts of prey) and prey (the animal victims of carnivores), and thus seems an unsuitable term. Vertebrate pests of crop plants are often termed herbivores. Insects noxious to plants are termed just that, or they are termed harmful insects. The parasite and the insect pest have their hosts, and the weed pest and the crop-plant victim are competitors, but there are no special terms for the crop-plant victims of vermin. Moreover, the interactions between plants and either vermin or noxious insects are seldom obligatory to the pest: the rat may consume garbage rather than stored corn; the white fly, though perhaps demonstrating a preference for *Viburnum*, will colonize, nevertheless, *Ligustrum*.

Plant-pest-control measures succeed when the manifold interactions between populations of pests and their victimized plants are somehow made favorable to the crop plant and unfavorable to its pests. In the final analysis, the successful pest-control measure inhibits the rise and hastens the fall of the pest population. The principles—fundamental truths—that undergird the structure of all effective plant-pest-control systems are treated next.

LITERATURE CITED

Boughey, A. S. 1971. *Fundamental Ecology.* San Francisco: Intext.

———. 1973. *Ecology of Populations.* New York: Macmillan.

Deevey, E. S. 1950. "The probability of death." *Scientific American,* April, 58–60.

Ehrlich, P. R., and A. H. Ehrlich. 1970. *Population, Resources, Environment.* San Francisco: W. H. Freeman.

Odum, E. P. 1971. *Fundamentals of Ecology,* 3rd ed. Philadelphia: Saunders.

Seinhorst, S. W. 1970. "Dynamics of populations of plant-parasitic nematodes." *Ann. Rev. Phytopathol.,* 8:131–156.

Zadoks, J. C. 1970. *Grondslagen van de Plantaartige Produktie Gewasbeschmerming, deel l, Diktaat.* Wageningen, The Netherlands: Laboratorium voor Fytopathologie.

5
Elements of Plant-Pest Control

The hundreds of specific measures used in efforts to control plant pests have been classified in several very different ways. For example, there are "biological, physical, chemical, and legal" methods of vertebrate-pest control (National Academy of Sciences, 1970); weed-control measures are either "eradication, control, or preventive weed control" (National Academy of Sciences, 1968b); nematodes are controlled by measures labeled "prevention of spread, cultural practices, biological resistance, physical, and chemical" (National Academy of Sciences, 1968c); plant-disease-control measures have been categorized as "exclusion, eradication, protection, resistance, therapy, and avoidance" (National Academy of Sciences, 1968a): and insects are controlled by measures classified as "regulatory measures, resistance, cultural control, microbial control, sterilization, and chemical control—including antimetabolites, hormones, deterrents, attractants, repellents, and insecticides" (National Academy of Sciences, 1969). The arrangement of all plant-pest-control measures into related groups would seem well-nigh impossible, considering the chaotic state of affairs suggested by the preceding enumeration. But the pest-control scientist must somehow make order out of this chaos in order to have a scientific basis on which plant-pest control can be integrated into a unified body of knowledge. It is the purpose of this chapter to classify measures of plant-pest control in terms of those fundamental laws we call principles.

POPULATION DYNAMICS AS THE BASIS
FOR PLANT-PEST CONTROL

Measures exercised in the attempts to control plant pests are, in essence, "moving forces behind fluctuations in . . .populations" of pests, and a successful control measure "inhibits the rise and hastens the fall of the pest population" (Chapter 4). Viewed scientifically, pest-control procedures should be classified in terms of how they prevent pest populations from reaching damaging sizes. In plant pathology, we can control the populations of pathogens in two ways. First, the amount of initial inoculum may be so reduced by sanitation, for example, that the inoculum never reaches agriculturally significant levels during the growing season of the crop plant. Second, the rate of increase in the amount of inoculum during the growing season can be retarded, by protective fungicides, for example. In practice, both objectives are sought in the same field before and throughout the same season. These approaches to plant-disease control, first stated by van der Plank (1968), are also valid approaches, in the broadest interpretation, to the control of all plant pests. We seek either to destroy the initial populations of plant pests or to decelerate the rates of increase in number of pests.

No plant pest would threaten our crops if its initial population were eliminated and if individuals were not introduced into the same area afterward. But total destruction of a plant-pest population is seldom achieved. Nevertheless, incomplete destruction of the initial population may lead to control, provided the remaining pests are too few in number to (1) inflict intolerable losses, and (2) initiate a "population explosion" before the harvest of the plant product. Control measures designed to reduce the numbers in initial populations are thus effective against plant pests with low "birth rates."

What of plant pests with high "birth rates"? These pests, notably such pathogens as plant-parasitic fungi, are characterized by a swift rise in the population when the biotic and abiotic factors of their surroundings are favorable to them. Although their initial populations may be so reduced that only a small percentage survives, that small percentage has such biotic potential that the pest population reaches destructive levels early in the cropping season. The only way to combat such pests successfully is to retard their rate of population increase in and among the crop plants we wish to protect.

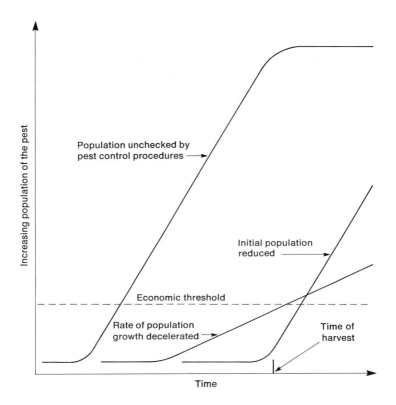

FIGURE   5.1

Plant-pest populations as affected by control practices that reduce the population initially or that decelerate its rate of increase.

We can visualize the end result of the two very different approaches to the control of plant-pest populations by a simple graph (Figure 5.1). A plant-pest population plotted against time yields a sigmoid curve, and at some point on that curve the pest population is said to have reached an intolerable level. Complete elimination of the initial population would "erase" the curve. If sufficient numbers of the initial population are killed and if the pest species has a low "birth rate," we can delay the burgeoning of its population. This would shift the curve to the right, and the destructive level of pest population would not be reached before the time of harvest. If we control plant pests, usually those with high "birth rates," by retarding the rate of increase in their numbers, we flatten the slope of the curve. This, too, could mean a bountiful harvest before pest populations have reached intolerable levels. The next question is: What are "intolerable" levels of plant-pest populations?

THE ECONOMIC THRESHOLD

The economic threshold is "the population [of a pest] that produces damage equal to the cost of preventing that damage" (Headley, 1972). Any pest population below the economic threshold is tolerable, any above, intolerable. The notion of economic threshold thus answers the question: "How much pest damage can we afford?" Or, perhaps more importantly, the concept of economic threshold is the basis for deciding if a certain pest-control procedure is worth the cost and effort. The cost of controlling a plant-pest population is relatively low when the pest population is low. But as the pest population increases, so does the cost of controlling it, and control costs increase at increasing rates. Finally, if we approach 100 percent pest control, the cost of obtaining only a slight increase in production becomes prohibitive. So pest populations at the economic threshold are optimal to the farmer, at least from the standpoint of economics (Headley, 1972).

The term economic threshold has heretofore been used mostly by economic entomologists, but there is no reason to restrict its meaning to insects. The concept is equally applicable to vermin, weeds, and plant pathogens, including plant-parasitic nematodes. The idea of using the economic threshold as a barometer to predict the need to exercise pest control is easy to grasp, but putting it to practice on the farm is difficult. In the first place, we do not always know when a plant-pest population will exceed the tolerable level; indeed, we do not often know for sure what that tolerable level is. There is, therefore, much need for continued research on assessing the nature and extent of the damage inflicted on crop plants by their pests as the growing season unfolds. Further, many pest-control measures, if they are to succeed, must be applied before the pest population reaches the economic threshold. This calls for continued studies on the prediction of disease epidemics and "population explosions" among insects, weeds, and vertebrate pests. Moreover, market values of plant products and costs of control measures may fluctuate during the cropping season. Hence the successful practitioner of plant-pest control must keep abreast of the economic factors that have a bearing on the economic threshold. Finally, control practices aimed at keeping a given plant-pest population at its economic threshold may significantly affect the population dynamics of other pests or even of beneficial organisms in the vicinity. The pest-control expert, then, must be cognizant of how his practices may affect the ecosystem and of how such effects may, in turn, bear upon the economic threshold of his "target" pest. The interactions of the many factors relating to population dynamics, economic thresholds, and control practices are so little understood that much additional research is called for; these interactions are so intertwined that the sophisticated technology of the systems approach (Chapter 11) and computer modeling and simulation are now called on in an effort to untangle this Gordian knot.

THE PRINCIPLES OF PLANT-PEST CONTROL

The goal of plant-pest control is to hold pest populations below their economic thresholds. As defined, pest control implies success. Although individual control measures may not be successful, they are at least attempts to achieve control. Such efforts may well reduce crop damage due to pests, but they do not come within the literal meaning of pest control unless they reduce pest populations below that "tolerable" or "acceptable" level defined in the preceding section as the economic threshold. To view the reduction in crop damage to some point *above* the economic threshold as being control is like describing a 25-foot broad jump, across a chasm 50 feet wide and 500 feet deep, as being successful.

The measures used to achieve plant-pest control can be divided into seven categories, all of which are defined later in this section. These categories are exclusion, eradication, therapy, vertical resistance, horizontal resistance, protection, and avoidance. Each category can be expressed in terms of principle. For example: A plant pest is controlled if the pest is excluded or eradicated; a plant pest is controlled if the victimized plant is "cured" by therapy, is made resistant by breeding and selection, is placed in an environment that makes avoidance possible, or is protected from the pest. To put it another way: there can be no intolerably severe pest damage if there has been successful application of any specific control measure that exemplifies any of the seven principles.

In practice, of course, several control measures, often representing several very different principles, are usually applied to a given crop plant in one location during a single growing season. Perhaps this multiplicity of pest-control procedures has lead to the term pest management, which is preferred by many to "pest control." This term erroneously suggests that people manage crop-plant pests when, in fact, all they manage are the pest-control procedures. Others prefer the term pest regulation. This term has some merit, for regulation implies control by principles or laws, such as those listed above. Nevertheless, I will use the term pest control to encompass all that one may do to keep pest populations below their economic thresholds. "Control" has been used for so long in this "generic" sense by scientist and layman alike that it seems inappropriate to suggest an alternative expression at this late date. Besides, another term for control would serve no useful purpose. Whatever one calls it, plant-pest control is achieved by effective applications of measures that illustrate the principles defined and discussed in the several sections that follow. The principles of plant-pest control can be divided into two groups: those that reduce pest populations; and those that decelerate their rates of increase.

Reduction of Populations of Plant Pests

Every plant-pest-control measure works by keeping pest populations below their economic thresholds, but some measures work directly against the individuals in the population, others indirectly, reducing the "birth rate" of those individuals. The test of whether a pest-control measure should be classed in one group or the other is to see what that measure does to the growth curve of the pest population (Figure 5.1). If a measure shifts the curve to the right in time, it has no effect on the slope of the curve and therefore exemplifies one of the several principles that are primarily aimed at the initial population of the pest. Measures that flatten the curve (those that reduce the value of its slope) exemplify the principles that are primarily aimed at slowing the rate of reproduction of pest populations.

The initial population of a plant pest may be reduced by any measure exemplifying any of four principles: exclusion, eradication, therapy, or vertical resistance. Exclusionary and eradicative measures are applied directly against the pest population. Measures representative of therapy and vertical resistance are applied to the potential or actual crop-plant victims, but they ultimately have the same effect as exclusionary and eradicative measures: they reduce the populations of the pests.

Exclusion

Exclusionary measures are those applied directly against plant pests to keep out those pests that are already outside: outside the field, the farm, the region, or the country. As used here, exclusion in no way implies elimination or expulsion, which are acts of eradication, treated in the next section. Legally enforced stoppages of plant pests at ports of entry—by inspection, interception, destruction, and quarantine—are perhaps the most familiar pest-control measures that exemplify the principle of exclusion. But other pest-control measures are also exclusionary. For example, planting seed that is certified to be free of weed seed, insects, and plant pathogens permits the planter to exclude pests that are not yet established on a farm. Exclusion as a term labeling a principle of pest control has been used very much as defined here by plant pathologists, less so by other pest-control scientists. Much of what the weed scientist terms preventive weed control, and what the entomologist and students of vertebrate-pest control term legal or regulatory practices, would be classed as exclusionary in the scheme presented here.

Exclusion depends not only on preventing the introduction of a plant pest into an area, however small, but also on having previously performed some act of control directly on the pest itself, in order to decrease its population in that area. By this reasoning, for example, use of a rat-proof storage bin for grain would not be exclusionary if nothing had been done to reduce the population of rats in the area. Instead, using a rat-proof bin is a treatment focused on the crop-plant product, and such a treatment, it seems, should be considered protective, as reiterated later in this chapter.

Eradication

Eradicative plant-pest-control measures are eliminative: they get rid of pests already present in an area. Eradicative measures are frequently taken against plant pests during the interim between growing seasons for crop plants, but some such measures may continue year round. Among the most familiar eradicative pest-control measures are: soil treatments with heat or chemicals to control weeds, insects, and plant pathogens; chemical treatments to poison vermin; dormant sprays with fungicides and insecticides; crop rotation with non-susceptible plant species to deny insects and pathogens their sustenance; and mechanical or chemical destruction of weeds.

The term eradication has been defined as complete elimination of a weed by weed scientists, and as elimination of an insect species by entomologists. The term is used here in a much broader sense, and would therefore include certain "control" practices of the weed scientist, "eradication" and "suppression" of the entomologist, and certain "legal" and chemical measures of the vermin-control specialist. The essence of an eradicative control measure is its reduction of pest populations to a point below economic thresholds by eliminative action taken directly against the pest itself.

Therapy

Plant-pest-control measures that exemplify all the other principles of the science are preventive, but therapeutic measures, as applied to plant diseases, are curative. Therapeutic measures come into play after the crop plant has been victimized to some extent by its pest. Therapeutic plant-pest-control measures, therefore, are defined as those that control plant pests by acting directly against the pests, but only after the pest-victim relationship has begun. Although therapy acts to reduce the population of the pest, therapeutic treatments, unlike exclusionary and eradicative ones, are made on the crop plant. For example, systemic fungicides and insecticides applied to plants act upon, respectively, infecting fungal parasites and colonizing insect pests. Also, heat therapy may be used to free propagative parts of plants of infecting pathogens, such as viruses, mycoplasmalike organisms, fungi, or nematodes. It is difficult to see how the principle of therapy might be used against weeds and vermin. True, removal of weeds after they have begun to compete with crop plants would constitute a control measure applied after the onset of a pest-victim relationship. But the treatment is eradicative, not therapeutic: it is made directly on the pest, not on the crop-plant victim.

The use of newly developed systemic fungicides in the therapeutic control (cure) of plant diseases has introduced a new problem for the plant-disease specialist: strains (pure lines, genetically) of plant-pathogenic fungi that are resistant to therapeutic chemicals are rapidly selected for in fungal populations exposed to those chemicals. The problem of resistance of target organisms to pesticidal chemicals has not yet plagued the weed scientist as seriously as it has the entomologist and the specialist in rodent control. The plant pathologist is already somewhat familiar with the problem; for more than two decades, he has seen the "miracle drugs"—antibiotics—fail to control bacterial plant diseases, because antibiotic-resistant strains of plant-pathogenic bacteria have quickly become prominent in the population of those organisms treated with antibiotics.

Why do pesticide-resistant strains of plant pests become so quickly the predominating strains of the population? The answer lies in the potency and metabolic specificity of the toxic action of the pesticide. If a potent pesticide acts specifically—that is, if it kills by blocking a single enzymatically catalyzed metabolic reaction—it exerts a selection pressure on the population of the pest that permits survival of only those strains that are resistant to the poison. The likelihood that such strains exist is high, because their resistance is derived from their capacity to bypass the metabolic "roadblock" imposed by the pes-

ticide and continue their metabolic activities via an alternative pathway. Because a single metabolic reaction is governed by a single gene,
a one-gene difference between an individual strain and its wild-type
sister strains could make the one strain resistant to a pesticide that has
a highly specific mode of toxic action.

The concept of pesticidal potency is also important, because, if the
pesticide is required in large amounts, the population of the pest
species might consist largely of strains that tolerate large doses of the
poison. Such weakly poisonous pesticides might not control plant
pests very effectively, but they would probably not select for
pesticide-resistant strains of the target organism rapidly.

Development of pesticide-resistant strains of plant pests is not an
effect unique to therapeutic chemicals. As long as the pesticide is
potent and specific in its mode of action, it must be expected to favor
the survival of resistant strains of the target organism. The chemical
may be therapeutic, but resistance to pesticides also occurs against
eradicative and protective chemicals. In fact, systemic fungicides and
insecticides act therapeutically when applied after attack by a pest,
protectively when applied before the pest arrives on the scene.

Vertical Resistance

Crop-plant resistance, in the broad sense, is the heritable capacity of
the plant species to withstand the onslaught of a would-be pest. Resistance is of two very different types: vertical and horizontal (van der
Plank, 1968). A vertically resistant crop-plant variety is almost completely resistant to the predominant strains of plant-pest species,
and that resistance is usually conferred by a single gene or a very few
genes. But a variety that is vertically resistant to one or more strains of
a plant-pest species may be susceptible to other strains of the same
pest species. Vertical resistance thus treats different strains differently; it is a kind of biological discrimination. Horizontal resistance,
discussed in the next section, treats all strains of a pest population in
about the same way; it means a degree of resistance to all strains.

Vertical resistance to a whole host of plant-pathogenic pests is now
commonplace among commercial varieties of crop plants. It is also
used against several devastating insects, notably the Hessian fly of
wheat, the phylloxora of grapevines, and the corn-leaf aphid (Day,
1972). But vertical resistance seems to play no role in the control of
weeds or vermin.

Vertical resistance carries with it certain advantages. First, it is easy
to recognize, for it is almost absolute, and is therefore relatively easy

to find in nature; it is most likely to occur in the region to which the plant-pest species and the crop-plant species (or its progenitors) are indigenous. Its effects also make vertical resistance readily apparent to the plant breeder who is testing the progenies of crossbred plant parents. Second, vertical resistance, because it is usually controlled by a single dominant gene, is relatively easy to incorporate into the genetic makeup of commercially acceptable crop-plant varieties.

Briefly, a program of breeding and selecting a crop-plant variety for vertical resistance to a pest would go something like this. Plants of a susceptible but otherwise horticulturally acceptable variety would be crossed with a wild type that bears the gene for resistance. All progeny of this cross would be resistant, because the gene for vertical resistance is dominant. Progenies from selfed individuals would then segregate, according to Mendel's laws, for resistance (homozygous and heterozygous) and for susceptibility. Plants grown from seed of resistant progenies could then be backcrossed with plants of the susceptible commercial variety. Again, resistant segregates would provide plants as parents for another backcross. Continued backcrossing for approximately seven generations would finally yield true-breeding progenies bearing the dominant gene for pest resistance. Thus a new pest-resistant variety is born.

But vertical resistance has one serious flaw: it stands up only against wild-type strains of the pest. And just as new chemical-resistant strains of pests arise, so do new strains emerge that can attack crop-plant varieties resistant to the common strains of the pest species. Vertical resistance can be viewed as acting in a highly specific manner, just like the potent, recently developed specific insecticides and fungicides. And a metabolic change in a function of a plant pathogen that is controlled by only a single gene may cause the resulting strain of the pathogen species to become virulent to the plant variety that is vertically resistant to the wild strains. So vertical resistance can be lost, just as effectiveness of antibiotics can be lost. How soon vertical resistance may be lost under field conditions varies with the pest species. It is lost soon, usually within two to five years, if the pest species has an obligate or near-obligate relation with the victim species. But vertical resistance may remain effective for a long time—perhaps more than 20 years—if the pest species is not obligate and if its wild-type strains, by saprophytic growth, can thus maintain their position of prominence in the population. Among plant-pathogenic pests, those whose annual life cycle includes an active stage of extensive saprophytic growth would be the most likely to maintain stable populations, with the predominating strains being those that are not virulent to the vertically resistant variety.

Decelerating the Rate
at Which Plant-Pest Populations Increase

The remaining three principles of plant-pest control work by keeping pest populations in check during the cropping seasons. Any measure that exemplifies horizontal resistance, protection, or avoidance seeks to maintain pest populations at sizes beneath their economic thresholds.

Horizontal Resistance

Whereas vertical resistance treats different strains of pest species very differently, horizontal resistance treats them all about the same. Vertical resistance, usually inherited monogenically, is absolute, or qualitative. But horizontal resistance, inherited polygenically, is relative, or quantitative. Vertical resistance brooks no infection by parasites, no infestation by insect pests; horizontal resistance permits some infection (colonization) by parasites, some infestation by insects. But horizontal resistance allows pest species only a fraction of what they need from victimized plants for their most rapid growth and reproduction.

Horizontal resistance has the advantage of stability: it does not wither under the pressure of attack by different strains of the pest species in different localities and over time. Horizontal resistance to plant pathogens may erode (Robinson, 1969), but only if those who are breeding and selecting for improved varieties neglect to test progenies of each cross for pest resistance. Why, then, isn't horizontal resistance more widely used? First, it is difficult to work with. It is expressed in degrees, and varies ever so gradually from questionable resistance, through moderate resistance, to high resistance. Because it is inherited polygenically, horizontal resistance occurs in every conceivable degree within the progenies of any cross. Moreover, many individual plants may by sheer accident escape attack by the pest of the moment in experimental plots. Such plants could then be classed as highly resistant when they are, in fact, highly susceptible. It thus becomes almost impossible to know which plants of a progeny to save for seed, which to discard. Second, the degree of horizontal resistance to a plant pest is often too low to ensure our keeping the population of the pest in question at or below its economic threshold.

Actually, horizontal resistance is more widely used than is commonly realized. We take it for granted, and give it scant notice. But for centuries, agriculturists have improved their crop-plant varieties by selection—by saving seed from only their best plants, for example. So horizontal resistance has been practiced for centuries. Moreover, most vertically resistant varieties also contain genes that confer horizontal resistance. This is important, for if vertical resistance should be lost in a given variety, its horizontal resistance would defend it, at least to some degree.

Like vertical resistance, horizontal resistance seems to have had its greatest impact in the control of disease-causing pests, plant pathogens. But horizontal resistance to some insects is manifested by the fact that some plant varieties are less attractive as hosts or victims to certain insects; for example, aphids prefer black rather than red raspberry plants. Finally, crop-plant varieties capable of competing successfully against weed pests could also be termed horizontally resistant: they slow the rate of growth and development of a weed pest; they inherit this competitive ability polygenically; and they treat all strains of the weed species in about the same way.

Protection

The term protection has been used in two very different ways by those engaged in the study and practice of plant-pest control. Some would have protection mean any effort that controls plant pests, but I prefer to use a more restrictive definition: protective measures of plant-pest control are those in which growing plants or plant products are so treated that a barrier—either chemical or physical—is placed on or immediately surrounding the plants to prevent pests in the vicinity from establishing pest-victim relations with them. Not only are protective measures applied to the plant, but they have the effect of slowing the rate of increase of the pest population. The few pests that can surmount a protective barrier are unlikely to increase above their economic thresholds before the growing plants mature or before their harvested products are used.

The principal problem created by the definition of protection used here, it seems, is that it becomes difficult to distinguish protection from exclusion and, possibly, from eradication. The distinctions are easy to grasp, however, when the object or "target" of the pest-control measure is identified. If we do something to the potential victim in order to control pests, the treatment is protective. If we attack the pest directly, the control measure is exclusionary or eradicative. Let us consider some examples. Inspection of ships and cargo to ensure their freedom from rats is exclusionary; killing rats with poison is eradicative; construction of rat-proof bins for the storage of corn is protective. Treatment of cereal-grain seed with pesticidal chemicals may be therapeutic (if there is a systemic effect), eradicative (from the viewpoint of the seedsman, if seed-contaminating fungal spores are killed), exclusionary (from the viewpoint of the farmer who buys pathogen-free seed), or even protective (if the chemical seed dressing guards the planted seed and young seedling against attack by seedling-blight fungi and seedling-attacking insects).

Perhaps the most familiar of all protective plant-pest-control measures are the applications of fungicidal, bactericidal, and insecticidal chemicals to growing crop plants. It is easy to see how these measures are, by definition, protective against plant-pathogenic bacteria and fungi. The relatively insoluble toxic residue on the protected surfaces of sprayed or dusted plants lies in wait for the pathogen to come to it. And then, just before their penetration through the plant surfaces, these pathogens are killed by the toxic action of the pesticide. The action of insecticidal sprays or dusts is somewhat different. There is certainly a "knock-down" eradicative action by insecticides as they are being applied to crop plants: many insect pests are killed directly by the lethal spray or dust. But there is also residual action by the insecticide, and this is surely protective. I view insecticidal sprays used in the fields as protective treatments, for three reasons: they are applied to the plants themselves; they reach only those insects already in association with crop plants; and, in the long run, they decelerate the rate of increase of the insect-pest populations in the field of plant-victim species.

Avoidance

Plant-pest-control measures in which people work with the environment to enable their plants to escape contact with pests exemplify the principle of avoidance. This seventh principle of plant-pest control is unique. Whereas the "target" of all other pest-control principles is either the pest or its victim, the "target" of avoidance is the third member of the triad, the environment. Avoidance is founded upon our recognition of the environment as a driving force behind many pest-victim interactions. Outstanding examples of avoidance are such cultural practices as late planting of corn or fall-sown wheat to avoid seedling blight, and the planting of fall-sown wheat after the "fly-free" date to avoid damage by the Hessian fly.

THE RELATIONSHIP OF PLANT-PEST CONTROL TO FUNCTIONAL LIFE HISTORIES OF CROP PLANTS

Crop plants are vulnerable to assault from their natural enemies, plant pests, at every stage of their development; so the defense of our crop plants must be a continuous effort if the victory is to be gained. Any of the vital functions of green plants (Chapter 3) might be impaired by any of the several kinds of pests enumerated in Chapter 1. The successful defender of crop plants must enforce the laws (principles) of pest control before, during, and after every cropping season.

First, consider the vital function of food storage in plant parts. Whether food materials in storage organs of plants are to be consumed by human beings or their animals, or whether they are to be conserved for the sprouting of plant-propagative parts the next season, stored food materials are often the object of attack by plant pests. Stored grain, for example, may be consumed by rats or weevils, or rotted by molds. Also, seed of crop plants may be contaminated by weed seed, which, although they do not interfere with the process of food storage, must be removed before planting time; otherwise they will give rise to weed plants that compete with seedlings of the crop-plant species. Sanitation practices (eradicative or exclusionary measures) are often effective against pests associated with storage organs of seed plants. The practices exemplifying the principle of avoidance (cold, dry storage rooms to prevent rot) are widely used. Also, the principle of protection may be used, as when grain is stored in rat-proof bins.

Second, plant-pest-control measures must be taken to prevent the loss of seedlings, which break down and use the food that had been stored in the seed. Eradicative measures, such as "sterilization" of soil by heat or chemicals, and exclusionary measures, such as planting pest-free seed, are important prior to planting. Protective control measures seem of greatest value immediately after planting. Growers may use seed dressings of fungicides and insecticides to protect seed and seedlings from damage inflicted by certain soil-borne fungi and by insects that prefer juvenile plants. And the use of alarms, repellents, and even scarecrows to frighten away bird pests may also be considered protective.

Third, the vital functions of plant roots, absorption of water and nutrients from the soil, must be safeguarded from root-infecting nematodes and other pathogens, from root-feeding insects that lurk in the soil, from burrowing vermin, and from competitive weeds. Pre-planting eradicative measures (chemical fumigation, crop rotation) can sometimes rid the soil of pests that attack the roots, and pre-planting exclusionary measures may be of value. Post-planting eradicative cultural practices are commonly used to control weed pests. Also, varietal resistance—vertical and horizontal—helps control many plant diseases that are caused by root-infecting fungi.

Fourth, the vital function of growth, which begins with meristematic activity, is adversely affected by plant pests. The effect may be direct, as when plant pathogens, such as smut fungi and many nematode species, invade meristematic areas. The effect on growth may be indirect, as when browsing animals feed on young, tender plant shoots. It would be foolhardy to generalize on the subject of controlling plant pests that affect meristematic activity. The success of measures depends on each pest and its life cycle and on each kind of crop-plant species. A few examples prove the point. First, all cereal smuts affect plant meristems, but a given smut may be controlled by resistance if it is localized, another by eradicative and exclusionary seed treatments if it contaminates grain saved for planting, and yet another by chemotherapeutic means if it has already infected the embryo of the seed. Second, many root-infecting nematodes may be controlled by fumigation of the soil, an eradicative measure. Third, the fungus-induced club root of cabbage may be controlled by liming the soil until its pH reaches a value of about 7.2. This cultural practice, which makes the environment unfavorable to germination of the spores of the parasitic fungal pest, exemplifies the principle of avoidance. Finally, a population of crop plants may be protected from browsing animals by fences.

Fifth, which pests adversely affect the vital processes of water conduction in the xylem, and how might such pests be controlled? Of all the kinds of plant pests, only a few attack the tissues of the xylem directly. These are the plant-parasitic wilt-inducing bacteria (including the rickettsiae) and fungi. Some sucking insects feed on the plant sap that flows in the xylem. It is not at all clear how much damage they cause by that act alone, because some species also inject a plant toxin while feeding. Sucking insects may also feed on the juices within the phloem, and the harm they do may result mainly from their phloem-feeding activities. But at least one xylem-feeding insect, the grape leafhopper, is the principal agent of transmission of the rickettsia whose infection of the xylem is lethal to grapevines. Vascular wilt diseases of herbaceous crop plants are usually controlled by the eradicative cultural practice of crop rotation and by vertical and horizontal varietal resistance. Such diseases of woody plants are sometimes prevented by controlling the insect pest that is also the agent of dispersal and inoculation. A case in point is the so-called Dutch elm disease. Elimination of all recently killed elmwood within about half a mile of all trees to be safeguarded has proved effective, provided such wood is destroyed early in the spring, before young adult bark beetles emerge contaminated with spores of the parasitic fungus. Also, spraying healthy elm trees in the spring with an appropriate insecticide will protect them from infestations by the fungus-carrying bark beetles.

Sixth, the vital function of photosynthesis must be protected from pest damage if a field of crop plants is to yield full measure. All kinds of plant pests can reduce photosynthetic activity. Generally speaking, protective control measures are the most effective ones against foliage-destroying plant pathogens and vertebrate animals. Weed pests growing among the maturing crop plants, of course, are controlled eradicatively. Finally, depending on the plant pathogen or harmful insect in question, varietal resistance may be the best control, especially against the highly specialized parasitic pathogens that cause foliage diseases of crop plants, such as cereal grains and meadow crops, which cannot be economically treated with protective fungicides.

Finally, the process of translocation of foodstuffs that have been elaborated by photosynthesis to distant storage organs may be directly affected in adverse ways by plant pathogens that parasitize the phloem or by insects that feed in the phloem. Of plant pathogens, only the viruses and the mycoplasmalike organisms (MLO's) parasitize the phloem, and they are transmitted by the only insects that feed exclusively on the phloem, the sucking insects of the orders Homoptera, Hemiptera, and Thysanoptera. As might be expected, protective insecticidal sprays and the use of predaceous insects are often effective in the control of phloem-feeding insects. Control of insect transmitters of MLO's and viruses that attack the phloem also helps control both pathogens. But the spread of many viruses is not checked by control of insect carriers. Most viruses of the mosaic type are acquired by aphids as they probe in the epidermal tissue of crop plants, and are transmitted in seconds to another plant. Consequently, many insects may transmit viruses in a sprayed field before they succumb to the toxic action of the insecticide. Moreover, not all transmitters are killed by the insecticide, and these can effect numerous transmissions in a short time. Most plant viruses and MLO's are controlled by destruction of weed hosts, by planting seed or propagative parts certified free of virus, and by using resistant crop-plant varieties. Since MLO's are vulnerable to antibiotics of the tetracycline type, therapeutic control also works well.

In conclusion, the control of plant pests rests upon the seven principles, whose representative control measures must be applied to safeguard the seven vital plant functions subject to attack by pests (Figure 5.2). Every scheme devised to control plant pests must consider procedures applicable before, after, and throughout the life history of the crop plant. Different kinds of plant pests and the methods used for their control are treated in Chapters 6–10 by specialists in the several disciplines that deal with the pests of crop plants.

The complexities of the science of plant-pest control, it seems, can best be mastered if its voluminous data are synthesized into an integrated whole by means of the systems approach, which is the subject

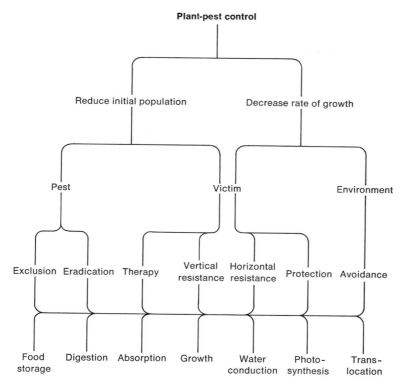

FIGURE   5.2

Interrelations between the seven principles of plant-pest control, the three "targets" of control practices, and the two approaches to keeping pests at or below their economic thresholds. Measures that exemplify the principles of plant-pest control safeguard one or more of seven vital functions performed by a crop plant during its life history.

of Chapter 11. How a fully integrated system of plant-pest control might serve the needs of world agriculture is treated in Chapter 12.

LITERATURE CITED

Day, P. R. 1972. *Genetics of Host-Parasite Interaction.* San Francisco: W. H. Freeman.

Headley, J. C. 1972. Defining the economic threshold In *Pest Control; Strategies for the Future.* Washington, DC: National Academy of Sciences.

_____. 1968a. *Plant Disease Development and Control.* Publ. 1596. Washington, DC.

_____. 1968b. *Weed Control.* Publ. 1597. Washington, DC.

_____. 1968c *Control of Plant-Parasitic Nematodes.* Publ. 1696. Washington, DC.

_____. 1969. *Insect-Pest Management and Control.* Publ. 1695. Washington, DC.

_____. 1970. *Vertebrate Pests: Problems and Control.* Washington, DC.

Robinson, R. S. 1969. "Disease resistance terminology." *Rev. Appl. Mycol.,* 48:593–606.

van der Plank, J. E. 1968. *Disease Resistance in Plants.* New York: Academic Press.

II

PLANT PESTS AND
THEIR CONTROL

6

Plant Pathology and the Control of Plant Diseases

Robert E. Stall and Daniel A. Roberts

Plant pathology is the study of disease in plants. Moreover, it is the science that deals not only with those continuous dysfunctions termed disease (effects), but also with parasitic plant pests (causes) and their control. The plant pathologist faces an unseen enemy, because almost all plant parasites are microscopic, and some, the viroids and viruses, are submicroscopic. Consequently, one who would control these pests of plants must first recognize and come to understand disease, the results of parasitic activity. The plant pathologist must also rely on assessments of disease intensity and severity as estimates of the magnitude of populations of those plant pests termed parasites. Comprehension of disease in plants is thus prerequisite to an understanding of the principles and procedures of controlling populations of plant parasites.

THE NATURE OF DISEASE IN PLANTS

Of all the plant damages wrought by pests, disease seems the least understood. Perhaps its mystery stems from the invisibility of its causes, perhaps from our ignorance of just exactly how a parasite nourishes itself deep within its living host. But several concepts emerge as the nature of plant disease is contemplated. These concepts, which have evolved during the last 150 years or so, have been expressed in recent books (Horsfall and Dimond, 1959–1960; National Academy of Sciences, 1968; Walker, 1969; Agrios, 1969; Roberts and Boothroyd, 1972; Kenaga, 1974) and are reiterated here to provide a basis for the definition of disease in plants.

Disease Is Harmful

Disease is harmful to the growth and reproduction of the green plant as an organism. Harmfulness to a few cells in a leaf, even though it might result in small but visible leafspots, does not amount to a disease, which affects the organism as a whole. In fact, insignificant flecks—indicative of the death of a few diseased cells—may be beneficial to the plant. Flecks are often marks of the hypersensitive defensive reaction characteristic of vertical resistance. Disease means harmfulness to the plant as a whole.

As it brings harm to the crop plant, disease also threatens the farmer with economic harm, even ruin. But monetary loss, the basis of our concept of the economic threshold, is not properly part of the concept of disease itself. The harmfulness of disease proper is harmfulness to the growth and reproduction of the crop plant in terms of its vital functions. With disease, as with all other pest damage, what is one farmer's economic loss may be another's windfall. Also, such natural events as the "bolting" of lettuce—the production of seed rather than edible heads—cannot be classed as disease, no matter how disastrous they are economically.

Disease Is Fundamentally Physiological

Disease adversely affects the behavior of crop plants. These functional disturbances also result in ugly form (morphological symptoms), but the harmful physiological effects are at the heart of our concept of disease in plants, so much so, in fact, that we classify specific diseases of plants on the basis of how they interrupt vital functions (McNew, 1950). Soft rots impair the storage of food, seedling blights impair digestion, root rots prevent absorption, tumors cause monstrous growth anomalies, vascular wilts impede the flow of water in diseased plants, leafspots (blights, mildews, and rusts) foul the photosynthetic machinery, and viral diseases prevent the translocation of foods elaborated by photosynthesis to their places of storage in fruit, seed, stems, and roots.

Disease Results from Continuous Irritation

Disease develops over time—in a single plant or in a population of plants in a field—and it develops continuously as the parasite irritates its host cells. It is this continuing irritation within the victimized plant that distinguishes disease caused by the parasitic pest from injury caused by all other plant pests, predaceous or competitive. The parasite is termed the primary causative agent of disease (the pathogen), to distinguish it from secondary agents, which are environmental factors favorable to disease.

The Consequences of Disease in Plants

Continuous irritation of a crop plant by its parasitic pest is the root cause of the harmfulness of disease. Parasitized cells are usually killed before they have completed their life expectancy. But parasitized cells do not always die quickly; the tissues they comprise are not visibly morbific at the outset. Sometimes they live, but do not divide and enlarge at a normal pace. Sometimes they live, but are mysteriously stimulated to divide and enlarge at many times their normal rates. Parasitized plants thus make three recognizable pathological responses: necrosis (death); hypoplasias (underdevelopments); and hyperplasias (overdevelopments) These responses, characteristic of a specific host-plant species that is parasitized by a specific pathogen, are termed symptoms. They are effects of disease, not disease itself or its cause. Because symptoms are characteristic of a given host-parasite interaction, they can be used to recognize or diagnose specific diseases of plants. Familiar diagnostic symptoms of disease in plants include: the necrotic symptoms of rot, canker, spot (Figure 6.1); blight, yellowing, and wilting (Figure 6.2); the hypoplastic symptoms of dwarfing, rosetting, mosaic (Figure 6.3); and chlorosis; and the hyperplastic symptoms of galls (Figure 6.4), blast, and witches broom.

A Definition of Disease in Plants

To regard disease in plants as continuous dysfunction is accurate, but incomplete; it does not accommodate all the concepts just considered. Perhaps the best synthesis of those concepts into a working definition of disease is that of Whetzel (1935). His definition is paraphrased here: in general, disease in plants is a series of harmful physiological processes caused by the continuous irritation of the plant by a primary agent, the pathogen; disease results in morbific cellular activity, and is expressed in characteristic pathological responses termed symptoms.

FIGURE 6.1

Blackspot disease of rose, illustrating the necrotic symptom, spot. [From *Fundamentals of Plant Pathology* by D. A. Roberts and C. W. Boothroyd. W. H. Freeman and Company. Copyright © 1972. Courtesy of Cornell University. Photograph by H. H. Lyon.]

PESTS THAT CAUSE PLANT DISEASES

The causative agents of disease in green plants number in the tens of thousands and include almost every form of life. But primary agents of disease may also be inanimate. Thus, nonliving (abiotic) agents of disease include mineral deficiencies and excesses, air pollutants, biologically produced toxicants, improperly used pesticidal chemicals, and such other environmental factors as wind, water, temperature, and sunlight. Nonliving things certainly qualify as primary agents of disease: they continuously irritate plant cells and tissues; they are harmful to the physiological processes of the plant; and they evoke pathological responses that manifest as the symptoms characteristic of the several diseases. But the abiotic agents of disease in

FIGURE   6.2

Dutch elm disease, illustrating the necrotic symptom of wilt. Left, healthy and diseased American elms. Right, transverse section through a diseased stem, showing darkening of the vascular tissues. [From *Fundamentals of Plant Pathology* by D. A. Roberts and C. W. Boothroyd. W. H. Freeman and Company. Copyright © 1972. Courtesy of Cornell University. Photograph by H. H. Lyon.]

FIGURE   6.3

Plant of Turkish tobacco infected by tobacco-mosaic virus. Light areas, in which chlorophyll never developed, illustrate the hypoplastic symptom of chlorosis. [Courtesy of Cornell University. Photograph by H. H. Lyon.]

FIGURE   6.4
Crown gall experimentally induced in the stem of a tomato plant, illustrating the hyperplastic symptom of tumefaction. [From *Fundamentals of Plant Pathology* by D. A. Roberts and C. W. Boothroyd. W. H. Freeman and Company. Copyright © 1972. Courtesy of Cornell University. Photograph by H. H. Lyon.]

plants, no matter how damaging, cannot be defined as plant pests, because plant pests are organisms noxious to plants. Because they cannot colonize plants and because they cannot be transmitted from plant to plant, the abiotic agents of plant disease are termed noninfectious, and the diseases they cause are termed noninfectious diseases. The important subject of noninfectious diseases is beyond the scope of this book, but it has been thoroughly treated elsewhere (Wallace, 1969; Sprague, 1964; Rich, 1964; Darley and Middleton, 1966; Treshow, 1971).

The primary agents of plant disease that qualify as pests are parasites, those living organisms that can colonize the tissues of their host-plant victims and can be transmitted from plant to plant. These biotic agents are, therefore, infectious, and the diseases they cause are termed infectious diseases. The infectious agents of plant diseases are treated in the standard textbooks on plant pathology. In this book, we will simply list the groups of living organisms of which some members are agents of disease in plants. The list is impressive: about 600 viruses and viroids, about 250 species of bacteria (including actinomycetes) and related organisms (mycoplasmalike organisms and rickettsiae), perhaps 20,000 species of fungi, possibly 1,000 species of nematodes (Chapter 7) and a few species each of algae, seed plants, protozoans, and arthropods (certain insects and spider mites). Little wonder that infectious plant diseases and the pests that cause them seem to defy understanding. Nevertheless, the complexities of causes and effects of disease in plants can be dealt with. We have already seen how plant diseases are classified in terms of sound concepts, and we will see, at the end of this chapter, how the procedures of plant-disease control exemplify the application of knowledge about the first principles of plant-pest control. At this point, let us examine the annual life cycles of plant pathogens as those cycles lead to disease. Regardless of its taxonomic position, every plant parasite develops in cycles of well-ordered events that proceed in sequence. As an example, the parasitic cycles of the rose-mildew fungus are diagramed in Figure 6.5.

First, the parasitic pest must produce an inoculum, some structure that is adapted for transmission to a healthy plant and that can either parasitize the host directly or develop another structure that can establish a parasitic relationship with the host. For example, inocula for viruses are the viral particles (virions); for bacteria, the bacterial cells; for fungi, various kinds of spores or the hyphal threads of mold; for nematodes, eggs or second-stage larvae.

Second, the inoculum must be transported from its source to a part of a host plant that can be infected. This dispersal of inoculum to suseptible tissue is termed inoculation. Agents of inoculation may be insects (for most viruses and mycoplasmalike organisms and for some bacteria and fungi), wind (for many fungi), and splashing rain (for many fungi).

Third, the parasite must enter the host plant, which it can do (depending on the organism) in one or more of three ways: through

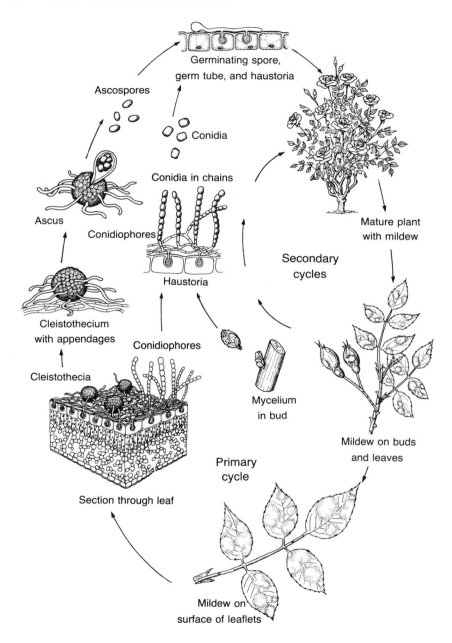

FIGURE   6.5

Diagram of the life cycle of *Sphaerotheca pannosa,* the fungus that causes powdery mildew of rose. [From *Fundamentals of Plant Pathology* by D. A. Roberts and C. W. Boothroyd. W. H. Freeman and Company. Copyright © 1972.]

wounds, through natural openings, or by growing directly through the unbroken protecting surface of the host. Viruses are literally injected into the plant as the homopterous insect carrier probes and feeds within its host. Bacteria depend on wounds or natural openings (for example, stomates, hydathodes, and lenticels) for entrance, but many fungi can penetrate plant parts by growing directly through plant surfaces, exerting enormous mechanical pressure and possibly softening host surfaces by enzymatic action.

Fourth, the parasite, now within its host, begins its parasitism. It obtains nourishment and grows within its host. This act of colonization is termed infection. The nature of parasitism still eludes us, but certainly the parasite damages the cytoplasmic membranes of the host cells, making those membranes freely permeable to solutes that would nourish the parasite. And parasitism certainly results from enzymatic attacks by the parasite upon carbohydrates, proteins, and lipids inside the host cell. The breakdown products of such complex molecules would diffuse across the damaged host-cell membranes and be absorbed by the parasite in the form of sugars, amino acids, and the like.

All such parasitic activity leads to the fifth and final event in the cycle: disease itself. Disease may be no more than the consequences of parasitism, or it may also result from the toxic action on the host of the parasite's metabolic by-products as it grows and colonizes its host.

The cyclic events just described are repeated slowly or rapidly, depending on the "natality" of the parasite. But unchecked, these cycles lead ultimately to a disease epidemic, a widespread outbreak of disease in a high percentage of the population of crop plants. Although the events leading to disease in a single cycle are always the same, life cycles of parasites are divided into two kinds, depending on what the pathogen was doing immediately before producing its inoculum. If the parasite was lying dormant because of season adversity before producing its inoculum, the cycle leading to disease is termed the primary cycle. Secondary cycles, sometimes called "summer cycles," begin with inocula produced by the parasite while it is actively colonizing its living host, the crop plant. Many plant parasites, usually the agents of foliar diseases, complete as many as twenty overlapping secondary cycles in a single cropping season. They have high "birth rates" and threaten to reach epidemic proportions within weeks. Others (for example, those causing root rots and vascular wilts) seldom complete secondary cycles. These soil-borne organisms have low "birth rates," and incite epidemics of plant diseases only after they have been developing for several years. But the epidemics they bring on slowly are ultimately just as devastating as the ones brought on swiftly by parasites with high "birth rates." Moreover, diseases caused by parasites that lurk in field soils are often among the most difficult to control, and pose a long-term threat equal in importance to that of the air-borne parasites of foliage, flowers, and fruit.

Comprehension of the nature of plant disease and knowledge of plant-parasitic pests and their life cycles are the bases for understanding disease in populations of crop plants. And disease in populations means epidemiology, the study of widespread outbreaks of disease.

EPIDEMIOLOGY OF PLANT DISEASES

If the science of epidemiology is to have practical value in the planning and executing of the procedures of plant-disease control, it must encompass more than an understanding of the progress of disease in a population of plants over time. It must also treat two other important subjects: the recognition of specific diseases (diagnosis) and the economic importance of specific plant diseases under various circumstances.

The Diagnosis of Plant Diseases

It is axiomatic that a disease must be identified before its economic effects and its rate of progress with time can be discovered. Recognizing and identifying plant diseases are accomplished by practicing the fine art of diagnosis. In terms of arts or skills, diagnosis means recognizing a disease from its symptoms. In terms of the science of plant pathology, diagnosis also means understanding the nature of a disease from its cause. Scrutiny of diseased plants thus reveals facts about causes and effects, and provides a sound basis for accurate decisions.

In practice, plant diseases in the field are usually diagnosed from symptoms. But symptoms are not always typical for a given disease. For example, the symptoms of several mosaic-viral diseases are striking and typical during cool weather, mild and almost invisible after prolonged hot spells. Moreover, symptoms of a specific disease might vary in severity and in appearance, depending on the strain of the parasite or the variety of the host plant. Finally, the symptoms of disease develop and change with time, making a complex syndrome or symptom picture. The art of disease diagnosis from symptoms can be developed only by experience, and the pest-control specialist can keep his diagnostic skills finely tuned only by constant practice in the field. With experience, the diagnostician learns to recognize certain unmistakable or key symptoms within each syndrome.

If the key symptom goes undetected, diagnosis from symptoms is unreliable, and the specialist then turns to an examination of diseased tissue in search of visible structures of the causative agent. Structures not easily seen in the field, even with the aid of a hand-held magnifying lens, may be detected microscopically in the laboratory. Disease diagnosis, then, usually relies on detecting key symptoms and distinctive structures of the causative agent. These structures are frequently termed "signs"of the pathogen.

Procedures somewhat more sophisticated than straightforward observations of symptoms and signs have been developed as aids to the diagnosis of many plant diseases. Serological methods and studies of insect-vector relationships often assist in the identification of plant viruses. The growth and appearance of colonies on different culture media may be used to distinguish species of plant-parasitic bacteria or fungi. Also, the results of chemical analyses of the tissue of diseased plants are sometimes indicative of infection by a given pathogen. Finally, experimentally established host ranges have proved valuable in diagnosis in the field; knowing which species and varieties are susceptible and which are insusceptible to a given parasite often makes its identification possible in an area where several different species and varieties of crop plants are commonly grown.

When diagnosis is not possible on the basis of symptoms and signs or on the basis of supplemental information, the plant pathologist can identify a plant pathogen by systematically testing four postulates. Three of these were stated by the nineteenth-century German bacteriologist Robert Koch, and the fourth was appended by Erwin F. Smith, the leading American student of plant-pathogenic bacteria during the first quarter of the twentieth century. The four rules for proving pathogenicity of a suspected causative agent of disease are universally known as Koch's Postulates, which state that:

1. The suspected agent must be constantly associated with a specific disease and its symptom syndrome.

2. The agent must be isolated from its host and be grown in pure culture.

3. Inoculum from that pure culture must infect host plants inoculated experimentally.

4. The suspected agent must be reisolated from experimentally diseased hosts, and its identity with the agent first isolated must be established.

Behind every field diagnosis based on symptoms is somewhere a published record proving fulfillment of Koch's Postulates—that is, for plant pathogens capable of saprophytic growth in culture. Proof of the pathogenicity of obligate parasites (those that can nourish themselves only from living hosts) rests on fulfillment of Koch's first and third Postulates.

Crop Losses to Plant Diseases

Major efforts of plant-disease control are directed toward those diseases that most drastically reduce the yields of the crop plants that provide us with food, clothing, and shelter. And this is fitting. But how do we decide where to make the greatest efforts? We must not only be able to recognize different diseases of important crop plants, but also be able to assess the magnitude of the losses due to each disease, if we would expend our limited amounts of time, energy, and money for the greatest good. Thus far, it seems, plant pathologists have not been very successful in discovering exactly how the incidence and severity of plant diseases—at different times throughout the growing season—affect the quantity and the quality of the desired product.

But some advances have been made, and the techniques for measuring crop losses due to disease are continually improving (Large, 1966). The incidence of most plant diseases can be measured, and some progress is being made in discovering the effects of plant diseases on yield and quality. Finally, plant pathologists are coming to grips with the complex question of how much disease can be tolerated at what time during the growing season and in what parts of the plant. For example, lower leaves of cereal plants contribute only 15 to 20 percent of the yield of grain; so considerable disease in those leaves is acceptable.

Estimates of crop losses to diseases are now based on the judgment of specialists in the field, and such estimates are probably accurate when made by an experienced specialist who is skilled as a keen observer. Such a specialist takes into account the percentage of diseased plants, the average severity of the disease in different parts of the plant, and the time of observation. Then, before recommending control action, he considers the increased cost of production due to the control procedure, weighing that added expense against the income it can be reasonably expected to generate. In effect, the appraiser of losses to plant diseases invokes the concept of the economic threshold.

The best estimates of crop losses to disease suggest that, on the average, diseases reap 15 to 20 percent of the most important food and fiber crops in the industrialized countries of the world. In the United States, for example, we lose to plant diseases more than 20 percent of our beans, wheat, potatoes, and tomatoes every year (LeClerg, 1964).

For all crop plants, however, diseases probably exact a toll amounting, on the average, to about 10 percent of the potential yields. We suffer whatever crop losses we do because we fail to prevent disease epidemics, which develop as the disease-progress curve ascends above the intolerable limits defined as economic thresholds.

The Progress of Plant Disease in Populations of Crop Plants

The plant pathologist seeks ultimately to prevent epidemics of plant diseases. His principal concerns, therefore, are populations of crop plants and populations of their parasites. He is concerned with the individuals within both populations only insofar as their behavior affects the development of epidemics. The specialist in plant-disease control thus deals mainly with population dynamics, and especially with the rise and fall of the populations of plant-parasitic pests. But a census of populations of parasites in the field is not possible, for parasites are the unseen enemy. We see only the effects of their destructive activities, and, by the time we notice such effects as symptoms of disease, the population of a plant parasite may have already passed its economic threshold. Nevertheless, the plant pathologist can assess disease incidence and severity, which bear a direct relation to the population of the parasite and to the rate at which that unwanted population increases in number. Even as populations of living things increase logarithmically to the carrying capacity of their environment, so the magnitude of disease severity in a field of crop plants increases logarithmically in time until a limit is reached (van der Plank, 1963). That limit is predestined by the numbers and susceptibility of the crop-plant variety, by the numbers and virulence of the parasite, and by the duration of environmental conditions favorable to the parasitic life cycles of the pathogens.

In practice, the disease-progress curve, the plant pathologist's substitute for a population-growth curve, is a plot of disease severity against time, and the characteristic sigmoid or logarithmic curve that invariably results is depicted in Figure 6.6. The datum used to denote disease severity along with the ordinate will be different for different kids of plant diseases. For example, it can be the percentage of the plants diseased when the effects are lethal, or it can be the percentage of necrotic tissues when the effects are not lethal. Also, the units of time along the abscissa differ with different diseases. For example, those units are measured in days for diseases caused by such parasites as the pathogen of southern leaf blight of corn (*Helminthosporium maydis*), a pathogen with an exceedingly high "birth rate." Time is

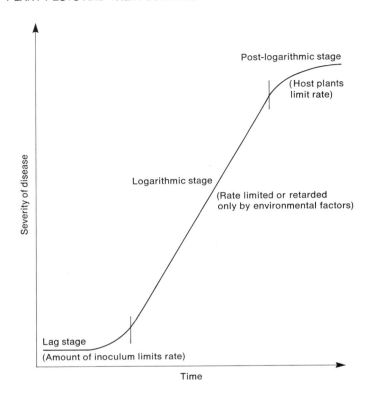

FIGURE   6.6

The disease-progress curve, which is an indirect assessment of the increase in numbers of individuals within a population of plant parasites.

expressed in months or years for diseases such as lethal yellowing of
coconut palms and foot rot of citrus, both caused by pathogens with
low "birth rates." In any event, the disease-progress curve is always
sigmoid, just like the growth curve for any population of organisms.

   The disease-progress curve develops in three separate stages: (1)
the lag stage; (2) the logarithmic stage; and (3) the stationary stage.
Factors that limit the rate of disease development are the relatively
low amounts of inoculum in the lag stage and the paucity of healthy
plants available to the inoculum in the stationary stage. The
logarithmic stage of the disease-progress curve provides a measure of
the so-called infection rate. It answers the question: How fast is the
epidemic developing? From this brief examination of the disease-

progress curve it becomes apparent that the rate at which epidemics of plant diseases progress depends on the magnitude of three fundamental variables: the initial inoculum (i), the infection rate (r), and time (t). Each acts as a limiting factor: if either i or r approaches zero, there can be no epidemic; if t is short, as when an early maturing variety is planted early, the disease-progress curve would never reach its theoretical stationary stage. If the disease-control specialist can use different control procedures, each of which affects a different factor among the three, he can usually achieve a synergistic effect. That is, the beneficial effects of the different control procedures combined are greater than would be expected from what each procedure might accomplish alone. Different control practices that affect only one variable, especially the rate of disease development, sometimes give a synergistic effect, but they usually have only an additive effect. For a given disease, then, the plant pathologist seeks to prevent an epidemic by adopting methods aimed at reducing the initial inoculum, the rate of infection, and the time of exposure of the crop plant to its parasitic pest. His actions against the initial inoculum would shift the disease-progress curve to the right, denoting a significant delay in the onset of the epidemic. His efforts at reducing the infection rate would so flatten the slope of the curve that the degree of disease severity would not exceed that point we know as the economic threshold. And his control practices might shorten the time available for the disease to reach epidemic levels.

PROCEDURES OF PLANT-DISEASE CONTROL

The fundamental laws or principles of plant-pest control have already been defined (Chapter 5). Here we will treat the procedures used by the practitioners of plant-disease control, and show how those procedures relate to epidemiology and how they exemplify the seven principles of plant-pest control. But first, two reminders: (1) all plant-pest control measures seek to reduce initial pest populations and to slow their rates of increase; and (2) any pest-control procedure is the practical application of one of seven principles, those of exclusion, eradication, therapy, vertical resistance, horizontal resistance, protection, and avoidance. The principles of plant-disease control are applied by the farmer by practice of three kinds of procedures: the use of disease-resistant varieties, the use of cultural practices, and the use of pesticidal chemicals.

Disease-Resistant Varieties

Disease resistance is the rule, susceptibility the exception. A single crop-plant species can be parasitized by only a fraction of 1 percent of the tens of thousands of those infectious organisms we call plant pathogens. The trouble is, any one of these relatively few parasitic pests can destroy any field of crop plants whenever conditions favorable to an epidemic persist. But the very epidemic that kills susceptible plants makes it possible to detect resistant individuals, which can exhibit their resistance only if exposed to parasitic attack. Plants that survive an epidemic—whether in the farmer's field or in the wild— possess desirable genetic traits that can be transferred to their progenies. Dedicated plant breeders, by their time-consuming and complicated programs of breeding and selection, have developed agricultural varieties of useful crop plants that resist the attacks of many important plant-parasitic pests. For the farmer, the use of disease-resistant varieties is often the easiest and most effective of all disease-control procedures.

But disease resistance poses numerous problems with which the pest-control specialist must deal. Behind the successful development and use of every disease-resistant variety lies the specialist's understanding of the sources of resistance in a plant species, the very different kinds of disease resistance, the methods of transferring the genes for resistance to obtain agriculturally acceptable resistant varieties, the genetic variability of the parasitic species, and the influence of environmental factors on the expression of resistance by the new variety. The remainder of this section is concerned with the sources of disease resistance in plants, the kinds of resistance, and the ways in which the different kinds of resistance prevent epidemics of plant diseases.

The Sources of Disease Resistance in Plants

There are only two sources of disease-resistant plants: agriculturally acceptable varieties of crop-plant species, and wild plants that can interbreed with agriculturally acceptable varieties. Resistant individuals, particularly within an agricultural variety, are first detected after an epidemic has run its course. But those plants that survive and are the least affected by disease do not always possess genes conferring resistance; they may have simply escaped inoculation. To test whether a disease-free survivor of an epidemic is truly resistant, seed from that plant are sown and the progeny exposed to the pathogen in question. Genes for resistance are indicated when the progeny prove more resistant than were the majority of the members of the preceding generation.

Resistant plants may be selected from a population after naturally occurring epidemics or after epidemics created under experimentally controlled conditions, in which test plants are inoculated with populations of the pathogen that have been standardized in quantity and quality. The experimentalist thus inoculates test plants (often under controlled environmental conditions) with measured amounts of inoculum obtained from a single strain of the pathogenic species. Such a strain, if distinguishable in terms of its pathogenicity to particular varieties of crop-plant species, is termed a pathological or physiological race. By exposing plants to inoculum prepared from several different races, the plant pathologist can measure precisely the reactions of the test plants to the genetic variants within the population of the parasitic species. This is not possible when naturally occurring epidemics are relied on, for the experimenter has no way of knowing whether the inoculum consisted of a single race or a mixture of races. Moreover, inoculum is not usually dispersed evenly over the field in naturally occurring epidemics. The scientist who would test for disease resistance under controlled conditions soon faces severe physical limitations: there's just not enough time, space, people, and money to permit adequate testing of many progenies (from several generations) against the onslaught of several different races of the parasitic pest. Consequently, naturally occurring epidemics are often relied on to test the disease reaction of plant populations. No matter how the resistant individuals are detected in a population of an agricultural variety, they serve as the progenitors of disease-resistant varieties that can be developed speedily. Open-pollinated varieties are usually better sources of resistance than are the homozygous clonal or self-pollinated ones, in which variants seldom occur.

Although the ancestors of disease-resistant varieties may be individuals of a crop-plant variety, wild plants are probably the most common sources of genes for disease resistance. The value of wild plants as sources of genes conferring disease resistance was first demonstrated by the eminent Russian geneticist, Nicolai Ivanovich Vavilov, whose untimely and tragic death in 1943 in a Soviet camp for political prisoners (Solzhenitsyn, 1975) was a severe blow to explorations in search of valuable plant-germ plasm.

The geographical center of origin of a wild-plant species is the most likely place to find plants with genes for resistance to disease— provided, of course, that the species of the parasite in question was also present in the same area. There, parasite and host would have evolved together for centuries, and the populations of the wild-plant species would have been subjected to repeated epidemics. Finally, only the disease-resistant individuals would remain. Sometimes a particular parasite does not occur in the center of origin of its host-plant species, but parasite and host may have existed together for a long time elsewhere, in the so-called secondary centers of origin. In any event, it is the exposure of host-plant species to parasitic-pest species—perhaps for thousands of years—that permits the natural selection of the fittest: those plants resistant to disease.

The plant-pest-control specialist, we believe, has the responsibility of discovering and exploring the centers of origin of plant species with the objective of making use of Nature's bounty. And scientists, in the finest tradition of Vavilov, have accepted that responsibility. Botanists explore the world over in search of desirable plants, and send seeds and other propagative parts of those plants to centers of collection and storage. For example, the Plant Introduction Section of the United States Department of Agriculture maintains a collection of seeds and other propagative plant parts, and special research groups test plants propagated from the collection for their resistance to important plant diseases.

The Kinds of Disease Resistance in Plants

Disease resistance in plants is of two very different kinds, vertical or horizontal (van der Plank, 1968). Some of the important characteristics of these two kinds of resistance are discussed in this section.

Vertical resistance has also been termed field immunity, type A resistance, or monogenic resistance. Because vertical resistance is absolute (there is no graded series of increasing or decreasing degrees of resistance), such resistance is also referred to as discrete resistance. A segregating progeny would consist of some highly resistant individuals, some totally susceptible ones. Vertical resistance is usually inherited monogenically, and resistance is almost always dominant over susceptibility. Resistance is usually expressed by the hypersensitive reaction. There is no disease in vertically resistant plants under normal field conditions, hence the term "field immunity." Vertical resistance is the same in seedlings and mature plants alike. Consequently, progenies can be reliably tested for resistance while still young.

The vulnerability of vertically resistant varieties to different races of a plant parasite was discussed in Chapter 5. Fortunately, vertical resistance works because populations of plant parasites in the field are made up of only a few races. It cannot be predicted with certainty how long a vertically resistant variety will remain so, for much depends on the frequencies at which new races arise by mutation or by hybridization, and on the aggressiveness and fitness for survival possessed by any new race that might arise.

The second kind of resistance, horizontal resistance, has also been termed field resistance, type B resistance, or polygenic resistance. Horizontal resistance is not absolute. For example, progenies of segregating generations would contain highly resistant individuals, highly susceptible ones, and some that are intermediate in susceptibility and resistance. Within the population, there would be a continuous grading of disease severity from one extreme to the other. This has led to the term "continuous resistance," used by some to denote horizontal resistance.

For a given disease, the number of genes in the plant necessary for its horizontal resistance determines the percentage of individuals (in a segregating population) whose degree of resistance equals that of the more resistant parent. Though polygenically inherited, horizontal resistance is conferred by only a fraction of 1 percent of the genes in the crop-plant species.

Horizontally resistant plants are seldom without some disease. They are, therefore, susceptible—at least from the scientific point of view. They are simply less susceptible than other plants. Symptoms of disease in horizontally resistant plants are identical to those in susceptible ones, but disease is less severe in resistant plants. That is, fewer plants are infected, and fewer lesions form in each infected plant. Greater amounts of inoculum may be needed to cause disease in horizontally resistant plants. Also, disease develops slowly in resistant plants, and the pathogen produces only small amounts of inoculum for secondary cycles during the cropping season. In short, horizontal resistance reduces the aggressiveness of plant-parasitic pests.

Horizontal resistance is not usually apparent in seedling plants, because even a few early infections, even though they enlarge slowly, can do much damage to juvenile plants. Consequently, testing progenies for horizontal resistance must be done in maturing plants, which require additional time and space. Also, amounts of initial inoculum so much affect the expression of horizontal resistance that the experimenter must often rely on naturally occurring epidemics to provide a reliable test for field resistance.

The expression of horizontal resistance under field conditions is much affected by external factors of the environment (Walker, 1969). Prominent among these are edaphic factors (soil fertility, soil moisture, soil temperature) and biotic factors (antibiosis between soil-borne parasites and saprophytes, presence of insect vectors of plant pathogens). An important example of a biotic factor that directly affects horizontal resistance is that wilt resistance in such plants as cotton and tomato is reduced if they are attacked by the root-knot nematode. The nematode predisposes its host-plant species to a vascular wilt disease.

Unlike vertical resistance, horizontal resistance is variable: disease in field-resistant plants may be mild one year, severe the next. But no matter how much disease resistance a horizontally resistant plant possesses, that resistance is expressed against all races of the parasite. True, some strains of a parasitic species are more aggressive than others, and cause somewhat more disease in horizontally resistant varieties. And, theoretically, natural selection for aggressiveness of a plant pathogen could, after a long time, bring on a loss of horizontal resistance. But there is no evidence yet that this is happening to any of our horizontally resistant crop-plant varieties.

The two kinds of disease resistance in plants, horizontal and vertical resistance, are distinguished graphically in Figure 6.7.

Disease Resistance in Farm Practice

The efforts expended by plant explorers on discovering sources of disease resistance, and the labors of the plant breeders and pathologists who incorporate genes for disease resistance into agriculturally acceptable crop-plant varieties, make the use of disease-resistant varieties perhaps the easiest disease control measure to put into practice on the farm. But the practicing farmer must recognize at least three circumstances that affect his successful practice of this easy-to-use disease-control measure.

First, the energies, time and money expended in the development of disease-resistant crop-plant varieties all aim toward producing a variety of a given plant species that is resistant to only one disease. The farmer must therefore use other control measures to combat other important diseases of this crop plant. For example, a variety of corn that is resistant to smut should have its seed treated with a fungicide to ensure protection against the seedling-blight disease.

Second, farmers who use a variety that is resistant to a particular disease must continually evaluate the performance of that variety in their own fields over the years. If they observe disease symptoms one year in plants of a vertically resistant variety, that variety should not be planted the next year. Disease in that variety is proof that a new and virulent race of the pathogen has blossomed forth on their farms. Conversely, appearance of some disease one year in plants of a horizontally resistant variety is no cause for alarm. The same variety can probably be used successfully the next year, when environmental fac-

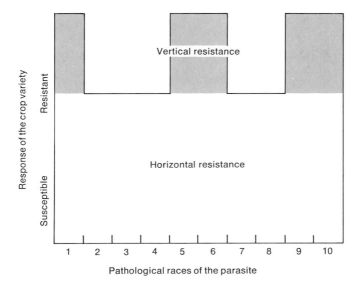

FIGURE   6.7

Graphic representation of the two kinds of resistance to plant diseases, vertical and horizontal. The hypothetical crop-plant variety is vertically resistant to hypothetical pathological races 1, 5, 6, 9, and 10. It is horizontally resistant to all races.

tors, which strongly affect horizontal resistance, will probably return to normal and will exert no particularly deleterious effect on the innate field resistance of the variety.

Third, when horizontally resistant varieties are used, the degree of resistance is often not sufficient to keep the population of the parasite below its economic threshold. Consequently, the disease to which a horizontally reistant variety is resistant may not be controlled by resistance alone. Resistance must be supplemented with other measures. For example, potato varieties horizontally resistant to late blight will probably have to be sprayed if the disease is to be controlled. But they will not have to be sprayed as often as those varieties that have no resistance. Moreover, since horizontal resistance decelerates the rate of infection, sanitary measures would have their greatest effects on late-blight control when the variety planted is horizontally resistant. In sum, development and use of disease-resistant varieties, though easy to apply, is a complex procedure for plant-disease control, and its complexities must be taken into account if we would practice that procedure effectively.

Cultural Practices

Cultural practices usually influence the development of disease in plants by affecting the environment. Such practices are intended to make the atmospheric, edaphic, or biological surroundings favorable to the crop plant, unfavorable to its parasites. The plant pathologist is chiefly concerned with the microclimate, that climate immediately around leaves, stems, and roots. Cultural practices that lead to disease control have little effect on the climate of a region, but can exert significant influence on the microclimate of the crop plants in a field.

The voluminous information on cultural practices in plant-disease control has already been reviewed (Horsfall and Dimond, 1959–1960). It is treated here in terms of how such practices affect the life cycles of parasites. Blocking a cycle at any point would achieve disease control. Three stages of a parasite's life cycle (survival between crops, production of inoculum for the primary cycle, and inoculation) are concerned with the initial inoculum, whereas the remaining stages (prepenetration and penetration, colonization and infection, and production of inoculum for secondary cycles) are concerned with the rate of disease development in the field.

Survival between Crops

Plant parasites survive between the cropping seasons of their hosts in a multitude of ways, but most overseason in one of four ways. First, some persist as parasites in perennial hosts or in stored or dormant plant parts. Second, some produce specialized resting structures that endure. Third, others survive as saprophytes in decaying crop residues in the soil. Fourth, some plant pathogens survive as residents within insects or within plant species that are not usually considered hosts. Plant-disease-control measures that prevent overseasoning are mostly eradicative.

Organisms that survive in the soil can often be controlled by crop rotations with insusceptible species. Depending on the system, either of two effects results. Usually the pathogen will produce inoculum, but will perish in the absence of a susceptible host. Sometimes, however, resistant crop plants used in the rotation stimulate inoculum production by parasites whose surviving structure might otherwise lie inert (in a resting condition) for years. Such crop plants used in systems of rotation are termed "catch crops," and have been used to control certain nematodes and other soil-borne pathogens.

Soil-borne plant pathogens can also be controlled by biological methods. Plant parasites may be controlled by antagonistic organisms that can be encouraged to grow luxuriantly by such cultural practices as green manuring and the use of appropriate soil additives. The soil-invading parasite thus becomes an amensal in association with its antagonist. The amensal effect may result from antibiosis, competition for substrate, predation, or hyperparasitism. The subject of biological control of plant diseases has been treated in considerable detail by Baker and Cook (1974).

Soil-borne plant parasites may also be killed during their over-seasoning stages by such cultural practices as deep plowing (as for the pathogen of southern leaf blight of corn), flooding (as for the cottony-rot pathogen and some nematodes), and frequent cultivation and fallow (as for the control of weeds that harbor plant viruses).

Plant diseases caused by organisms that survive as parasites within perennial hosts or within the seed of annual plants may be controlled therapeutically. Therapeutic treatments of heat and surgery are applicable here; those involving the use of chemicals will be mentioned later. Removal of cankered limbs (surgery) helps control fire blight of pears, and the hot-water treatment of cabbage seed controls the bacterial disease known as black rot. Heat therapy is also used to rid perennial hosts of plant-parasitic nematodes.

Production of Inoculum for the Primary Cycle

Whereas certain cultural practices have been used successfully to prevent the overseasoning of plant pathogens, they are of little value in preventing the production of inocula by organisms that do survive between cropping seasons. Nevertheless, environmental factors (particularly temperature, water, and organic and inorganic nutrients) significantly affect inoculum production. Warm temperature usually breaks dormancy of overseasoning structures; rain may leach growth inhibitors from the soil and permit germination of resting spores; and special nutrients may stimulate the growth of overseasoning structures that produce inoculum. Cultural practices that would alter (to the detriment of parasites and to the benefit of host plants) the temperature, water, or nutrient relations of parasites and hosts could lead to disease control by preventing inoculum production.

Dispersal of Inoculum and Inoculation

Two very different kinds of dispersal account for the spread of plant-parasitic pests from one place to another. First, there is the distribution of overseasoning structures (or inocula) into a geographical area never before occupied by the parasite. Human beings are the principal agents of such distribution, but insects, birds, and wind currents may also be involved. Exclusionary practices are most effective against human distribution. Second, there is the local dissemination of inoculum (for primary or secondary cycles of the pathogen), usually by insect vectors or by wind, flood waters, or splashing rain. Cultural practices that exemplify avoidance are sometimes used to prevent effective dissemination.

Because people are usually the agents who distribute plant pathogens into new areas, it is only fitting that we have devised regulations that seek to prevent distribution. Regulatory activities that involve the farmer directly include programs for the certification of propagative materials as being free of plant pathogens. Certification programs are voluntary, and are usually initiated by agricultural leaders who wish to maintain quality propagative stock and to provide farmers with such stock. State agencies usually enforce the standards of the program, and the large costs of obtaining disease-free "parent" plants, inspecting the candidate-seed fields, and administering the certificaton program are shared by participating farmers. Certification does not imply that the propagative plant parts are free from all diseases, or that they are completely free from a specific disease, because levels of tolerance are commonly established for specific diseases.

A second hierarchy of regulatory disease control is plant quarantine, the legally enforced stoppage of plant pathogens at points of entry into political subdivisions. The farmer is only indirectly involved in quarantines, but these regulations prevent distribution of plant pathogens and, like certification programs, provide the farmer with pathogen-free propagative materials. The Plant Quarantine Act of the United States governs importation of plant materials into the country and requires the state governments to enforce particular measures. Also, states make regulations concerning the movement of plant materials into them or within them. For example, Florida imposes a quarantine against the citrus-canker bacterium, which was eliminated from the state earlier by means of cooperative efforts led by the Florida Department of Agriculture.

Few farm-management practices can control plant diseases by preventing local dissemination during primary and secondary cycles of a pathogen. But the farmer can take advantage of environmental factors that influence dissemination and practice methods of avoidance aimed at lessening the chances of inoculation. For example, when several contiguous fields are planted with the same crop at intervals of time, fields to the leeward should be planted first, so that prevailing winds would disseminate air-borne inocula from the oldest fields (where it would develop first) away from the fields planted later.

Sometimes, relatively wide spacing of plants in a row permits plants to avoid inoculation. Less leafspot occurs in sugar beets that are spaced widely, because spores splashed by rain or wafted by air currents are less likely to be transported from diseased to healthy plants.

There is an unusual cultural practice that seems effective in preventing inoculation of plants by pathogens, especially viruses, disseminated by insect vectors: the use of barrier crops (border crops). For example, 50- to 100-foot strips of corn planted around a field of peppers offer a barrier to viruliferous aphids that might fly into the field after having fed on virus-infected weeds nearby. The aphids would be "trapped" in the virus-resistant corn plants, which they would colonize, and would be vulnerable to insecticides applied to the corn. Moreover, the aphids would lose stylet-borne (nonpersistent) viruses soon after colonizing the corn. Barrier crops would prevent inoculation and their use would exemplify the principle of avoidance. Barrier crops delay the onset of an epidemic, and the farmer contemplating their use must decide whether that delay is likely to increase yield and profit enough to counteract the obvious inconvenience and expense of the barrier crop.

Penetration and Infection Phenomena

These terminal events in the pathogenic life cycles of plant parasites are most frequently interfered with by the use of disease-resistant varieties and by chemical control methods involving protection or therapy. But the farmer can also use several cultural procedures that prevent extensive penetration and infection. Perhaps the most obvious is to prevent the wounding of plants by insects or farm machinery. Pathogens that require wounds for entry into their hosts cannot breach the outer coverings of undamaged plants.

The time required for penetration and infection varies with the kind of parasite. These events are much telescoped in time when inocula of viroids, viruses, mycoplasmalike organisms, or rickettsiae are literally injected into their hosts by the feeding activities of their insect vectors. Inoculation, penetration, and infection are almost simultaneous. Bacteria take longer, but will soon perish on the surface of host plants if they cannot enter through wounds or natural openings. Nematodes quickly batter their ways into host cells and tissues by the action of their spear-like stylets, and require very little time for penetration. But fungi, particularly those that invade foliage, require hours to complete the penetration and infection stages of their life cycles. Typically, a leaf-infecting fungus produces inoculum in the form of a spore, which is deposited on the leaf after being dispersed by wind or splashing rain. The spore must germinate, develop a thread-like germ tube, have its tip adhere to the leaf surface, and finally send forth a branch that grows into the plant, penetrating directly or through an open stomate (depending on the species of the fungus). A food relationship is established between parasite and host, and colonization is underway.

Everything that happens between inoculation and colonization is much affected by environmental factors, particularly water, but also temperature and nutrients on the surfaces of leaves. Except for the powdery mildew fungi (whose spores germinate best at high relative humidities, hardly at all in liquid water), spores of leaf-infecting fungi require a film of free (liquid) water for their germination. Also, every fungus has an optimum temperature for its growth and development. Consequently, the longer the leaves of the host plant remain wet and the longer the temperature remains near the optimum, the greater the chance of an epidemic.

Although we can do nothing about the conditions of moisture and temperature prevailing in the field, we can, by means of some cultural practices, affect the microclimate. Dew forms at night, and the length of time for which leaves remain wet the next day may be too short to permit germination of all fungal spores. This would mean relatively slow rates of disease development. But if crop plants are irrigated by sprinklers early in the morning, the rate of disease development will climb sharply and in direct proportion to the time that free water remains on the foliage. Irrigation not begun until the dew has evaporated breaks the period of wetting, and therefore, also breaks the infection cycle. Farmers can space plants widely and orient crop rows along a north-south axis to hasten the drying of foliage, thereby decelerating the rate of an epidemic of a fungal blight or leafspot disease.

Post-harvest diseases can be controlled by practices that prevent the extensive colonization of plant tissues after initial infection. Vegetables and fruit in transit to market and in storage are kept as cold and as dry as possible without damaging their quality.

All cultural practices that break the pathogenic cycles of plant parasites take advantage of environmental conditions that are unfavorable to disease development. Although such practices as we have mentioned obviously reduce initial inocula or reduce infection rates, the question often remains to plague us: is the cost of the practice less than anticipated returns? Recent advances in the technology of the systems approach to plant-pest control (Chapter 11) may show us how to calculate economic thresholds. The simulation of plant-disease epidemics by computer programming permits precise assessment of how a break in the life cycle of the pathogen will affect the total amount of disease at the end of the cropping season. Cost-benefit ratios obtained from models of epidemics would permit evaluation of cultural practices recommended for disease control. Reliable cost-benefit ratios would be equally applicable to disease control by chemicals.

Chemical Control

The pesticidal chemicals that control plant diseases may be used in very different ways, depending on the parasite to be controlled and on the circumstances it requires for parasitic activities. For example, a water-soluble eradicative spray is applied once to dormant peach trees to rid them of the overwintering spores of the fungus of peach-leaf curl, whereas relatively insoluble protective fungicides are applied repeatedly to the green leaves of potato plants to safeguard them from penetration by the fungus of late blight. Also, systemic fungicidal chemicals may be used therapeutically. The oxathiin derivatives that kill the smut fungi that infect embryos are therapeutic, as is benomyl (which has systemic action against powdery mildews and other leaf-infecting fungi). Volatile fungicides are often useful as soil-fumigating chemicals that have eradicative action. The nature of fungicidal chemicals has been discussed by Horsfall (1956), Torgeson (1967-1968), and Sharvelle (1961), and descriptions, uses, and modes of action of important fungicides and bactericides have been discussed by Ware (1975), who also covers other kinds of useful pesticides. In this section, we treat the chemical control of plant diseases in three categories: seed treatments, soil treatments, and protective sprays and dusts. Some of the chemicals currently recommended for these purposes are listed in Table 6.1.

Seed Treatments

Chemical treatments of seed may be effective in controlling plant pathogens in, on, and around planted seed. Seed treatment is therapeutic when it kills bacteria or fungi that infect embryos, cotyledons, or endosperms under the seed coat, eradicative when it kills spores of fungi that contaminate seed surfaces, and protective when it prevents penetration of soil-borne fungi into seedling stems. Also, seed treatments may be viewed as exclusionary, for the farmer usually purchases seed that is already freed of pathogens by chemical treatments, and thus excludes from the fields those organisms that over-season by infecting or infesting the seed.

TABLE 6.1.

Some chemicals commonly used to control plant diseases.[a]

Chemical and use	Relative toxicity	
	Oral	Dermal
Seed treatments (all fungicides)		
Chloraneb	Low	Low
Dichlone	Low	High
Thiram	Moderate	High
Carboxin (systemic and therapeutic)	Low	Low
Soil treatments		
Methyl bromide[b] (general pesticide)	Very high	Very high
PCNB (fungicide)	Low	Moderate
SMDC [vapam] (fungicide, nematicide)	Moderate	Moderate
MIT ["Vorlex"] (fungicide, nematicide)	Moderate	Moderate
D-D mixture (nematicide)	Moderate	Low
Plant-protective treatments		
Copper compounds (fungicides, bactericides)	Moderate	Low
Sulfur (fungicide)	Low	Moderate
Maneb (fungicide)	Very low	Low
Zineb (fungicide)	Very low	Low
Captan (fungicide)	Very low	Very low
Dinocap (fungicide for powdery mildews)	Low	Low
Streptomycin (bactericidal antibiotic)	Very low	Low
Cyclohexamide[b] (fungicidal antibiotic)	Very high	Very high
Benomyl (protective and therapeutic fungicide)	Very low	Very low

[a]*Source:* Pesticide Conference, Ithaca, NY, November 8–11, 1971.

[b]Dangerous pesticide; permit from the appropriate regulatory agency usually required for use.

Whereas the oxathiins (carboxin, DMOC) used to kill embryo-infecting smuts of cereal grains have little effect on other organisms, most eradicative and protective chemicals have a wide range of fungicidal activity; they are effective against most seed-infesting and seedling-blight fungi. But specific seed-treatment chemicals often work best to control a given disease of a single crop-plant species. Moreover, the toxicity of chemicals to seeds varies, and farmers should use only the compounds recommended by the Cooperative Extension Service of their country and state.

Copper- and mercury-containing compounds were first used as seed-treating chemicals. But copper is toxic to most seeds and seedlings, and mercury has been banned from use in seed treatments because of the danger it poses to humans and animals. Organic compounds now widely used as protective and eradicative seed treatments include thiram, chloraneb, dichlone, dexon, and captan.

Soil Treatments

Soil-borne plant pathogens greatly increase their populations as soils are cropped continuously, and finally reach such levels that contaminated soils are unfit for crop production. Chemical treatments of soil that eradicate the plant pathogens therein offer the opportunity of rapid reclamation of infested soils for agricultural uses. It is not uncommon to recommend preplanting chemical treatment of field soils for the control of nematode-induced diseases, and fumigation of seedbed and greenhouse soils (with methyl bromide, for example) is commonly practiced to eradicate weeds, insects, and plant pathogens. Field applications of soil-treatment chemicals for fungus control are usually restricted to treatments of furrows. Formaldehyde or captan applied in the row at seeding time will control seedling infection of onions by the smut fungus; pentachloronitrobenzene (PCNB) so applied is effective against sclerotia-producing fungi that cause seedling blights, stem rots, and root rots of many field crops. But soil-treatment fungicides are sometimes applied broadcast: turfgrasses, for example, may be drenched with benomyl or truban for the control of soil-borne fungi. Other soil-treatment fungicides are vapam and "Vorlex."

If a volatile chemical (methyl bromide) is recommended as a soil treatment, the farmer should realize that the treatment must be made one to two weeks before planting and that the treated soil should be "sealed" by mechanical compaction, with water, or by a mulch of some kind. If the chemical recommended is relatively nonvolatile ("Vorlex"), the farmer must be prepared to mix the material with the soil, as by disking or rototilling. Because nonvolatile soil-treatment chemicals are not phytotoxic, they can be incorporated into the soil by means of the planting operations. Soil treatments made at the time of planting are most effective against parasitic attacks that come early in the growing season.

Chemical Protection of Growing Plants

Growing plants, so vulnerable to the blights and mildews caused by "high-birth-rate" pathogens, must often be protected by topical application of fungicidal and bactericidal sprays or dusts. Protective fungicides, unlike the eradicative ones, are relatively insoluble in water and adhere to the surfaces to which they are applied. They cannot enter protected foliage in large amounts; instead, protective fungicides lie in wait for inocula, and slow the rate of disease development as they attack inocula on protected surfaces. When sprayed leaf surfaces are wetted, small amounts of the toxic ion or radical from the fungicidal residue are dissolved. The inocula (usually fungal spores) then absorb and accumulate the toxicant as they imbibe water for germination. Germination, growth, and penetration are thus prevented by protective fungicides.

Protective fungicides have undergone three stages in their development: first, inorganic compounds of copper and sulfur were used exclusively; second, complex organic materials began to replace the more phytotoxic inorganic fungicides; and, third, protective fungicides with systemic (therapeutic) action appeared in the 1960's and 1970's. Fungicides in the first two groups, though less toxic, have wider spectra of activity than those in the third group. But the numbers of parasitic species controlled by different fungicides may vary widely from fungicide to fungicide. For example, dodine is effective against apple scab, not so effective against other diseases of foliage; benomyl is effective against nonresistant strains (see Chapter 5) of such parasites as powdery mildews and other Ascomycetes, ineffective against most Phycomycetes. The farmer must select the proper fungicide, according to the published recommendation of the local Cooperative Extension Service. Some of the most frequently recommended protective fungicides include compounds of copper, sulfur, ferbam, maneb, zineb, captan, dinocap, carboxin, and benomyl. New fungicides appear on the scene from time to time, and our list of commonly used protective fungicides may change drastically by the beginning of the next century.

In order to use protective fungicides effectively, the farmer must not only select the right fungicide for the job, but also apply it in the right amount, at the right times, and in the right way. Too little fungicide fails to control disease; too much may be toxic to the plants to be protected. The farmer and applicator, therefore, must always follow use instructions to the letter. Timing of applications is also critical. For example, a spray applied one day too early or one day too late may fail to control northern leaf blight of corn (Berger, 1973). The frequency of protective spraying needed will also vary with the "weathering" properties of the fungicide used and with the rate of plant growth. After all, the new growth remains vulnerable until protected. The frequency of fungicidal applications can be adjusted to changing weather. Environmental factors affect the rate of foliage-disease development so strongly that fewer sprays are needed when weather is unfavorable to disease. Finally, the method of fungicide application is important, and the critical point is to achieve maximum coverage of susceptible foliage. The practicing farmer, therefore, must take into account such items as the volume of spray to use per unit of area, the shape and direction of the blast from the sprayer, the rate of speed of the tractor that moves the spray machine, the orifice size of the spray nozzles, the pressure in the spray tank, and so on.

In conclusion, a few words of caution: treat all pesticidal chemicals as if they were deadly poisons. Of course, most protective fungicides are thousands of times less toxic to human beings than some of the potent rodenticides or insecticides. But farm workers must handle all kinds of chemicals, and their supervisors should insist on impeccable deportment in the presence of all farm chemicals: don't spill them; don't inhale them; don't swallow them. Be prepared for quick treatment of accident victims. "People who use agricultural chemicals should handle them carefully, measure them accurately, and apply them correctly" (Roberts and Boothroyd, 1972).

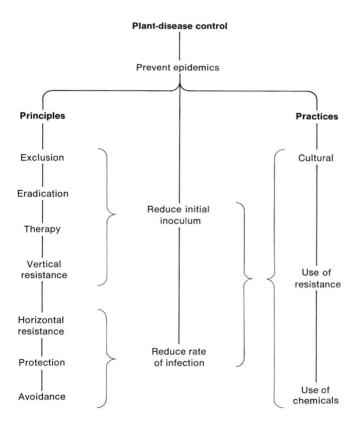

FIGURE   6.8

Interrelationships between the principles and practices of plant-disease control in terms of the bases for preventing epidemics of plant diseases.

A Network of Principles and Practices
of Plant-Disease Control

Every plant-disease-control measure should be applied by the farmer within the framework of the procedures described in this chapter. And, as we have seen, every plant-disease-control measure prevents completion of a particular event in the pathogenic life cycle of the target organism, the plant parasite. We have also seen, in this chapter and in Chapters 4 and 5, that every plant-disease-control measure exemplifies a fundamental law (principle) of our science, and that every principle affects either the initial inoculum or the infection rate. Further, a given plant-disease-control measure is applied directly to the pathogen, directly to the crop-plant species (and therefore indirectly against the pathogen), or to the environment (to the benefit of the host, to the hurt of the parasite). How the procedures used by farmers to combat plant diseases relate to the foundations of plant-disease control is depicted in Figure 6.8. Practices and principles alike focus on the two fundamental approaches to the prevention of epidemics: reducing the initial inoculum and decelerating the rate of infection. We believe the practices of all plant-pest control—not just those aimed at plant-disease control—are based on the first principles of our science.

LITERATURE CITED

Agrios, G. N. 1969. *Plant Pathology.* New York: Academic Press.

Baker, K. F., and R. J. Cook. 1974. *Biological Control of Plant Pathogens.* San Francisco: W. H. Freeman.

Berger, R. D. 1973. "*Helminthosporium turcicum* lesion numbers related to numbers of trapped spores and fungicide sprays." *Phytopathology,* 63:930–933.

Darley, E. F., and J. T. Middleton. 1966. "Problems of air pollution in plant pathology." *Ann. Rev. Phytopathol.,* 4:113–118.

Horsfall, J. G. 1956. *Principles of Fungicidal Action.* Waltham, MA: Chronica Botanica.

_____, and A.E. Dimond, eds. 1959–1960. *Plant Pathology: An Advanced Treatise,* 3 vols. New York: Academic Press.

Kenaga, C. B. 1974. *Principles of Phytopathology,* 2nd ed. Lafayette, IN: Balt Publishing Co.

Large, E. 1966. "Measuring plant disease." *Ann. Rev. Phytopathol.,* 4:9–28.

LeClerg, E. L. 1964. "Crop losses due to plant diseases in the United States." *Phytopathology.* 54:1309–1313.

McNew, G. L. 1950. "Outline of a new approach in teaching plant pathology." *Plant Dis. Rep.*, 34:106–110.

National Academy of Sciences. 1968. *Plant Disease Development and Control.* Publ. 1596. Washington, DC.

Rich, S. 1964. "Ozone damage to plants." *Ann. Rev. Phytopathol.*, 2:253–266.

Roberts, D. A., and C. W. Boothroyd. 1972. *Fundamentals of Plant Pathology.* San Francisco: W. H. Freeman.

Sharvelle, E. G. 1961. *The Nature and Uses of Modern Fungicides.* Minneapolis: Burgess.

Solzhenitsyn, A. I. 1975. *The Gulag Archipelago*, III–IV. New York: Harper & Row.

Sprague, H. B., ed. 1964. *Hunger Signs in Crops*, 3rd ed. New York: David McKay.

Torgeson, D. C., ed. 1967–1968. *Fungicides: An Advanced Treatise*, 2 vols. New York: Academic Press.

Treshow, M. 1971. "Fluorides as air pollutants affecting plants." *Ann. Rev. Phytopathol.*, 9:21–44.

van der Plank, J. E. 1963. *Plant Diseases: Epidemics and Control.* New York: Academic Press.

———, 1968. *Disease Resistance in Plants.* New York: Academic Press.

Walker, J. C. 1969. *Plant Pathology*, 3rd ed. New York: McGraw-Hill.

Wallace, T. 1961. *The Diagnosis of Mineral Deficiencies in Plants.* New York: Chemical Publishing.

Ware, G. W. 1975. *Pesticides: An Auto-Tutorial Approach.* San Francisco: W. H. Freeman.

Whetzel, H. H. 1935. "Nature of disease in plants." Mimeographed paper, Department of Plant Pathology, Cornell University, Ithaca, NY.

7

Plant-Parasitic Nematodes and Their Control

Vernon G. Perry

Nematodes are among the most numerous forms of multicellular animals found on earth. They inhabit soil, fresh water, and salt water. Most species are beneficial to the environment and to human beings because they feed upon organic matter and help break it down into simple compounds. Nematodes, in turn, are consumed by predaceous insects and other nematodes and are destroyed by parasitic fungi. Nematodes thus serve as important links in food chains. Few species are parasitic to animals and plants, in which they also induce disease. But the pathogenic forms have been studied most, and are the ones for which control measures have been developed. Some important plant-pathogenic nematodes are enumerated in Table 7.1.

Many kinds of nematodes, saprophytic as well as parasitic, live in the same environment with plants (Figure 7.1). Soil-inhabiting forms are most numerous near plant roots, which most plant-parasitic species feed on. These small parasites, seldom longer than two millimeters, are fundamentally aquatic, and live part or most of their lives in the water films that surround soil particles (Chitwood and Chitwood, 1950). A few species can swim up external films of moisture to attack buds and foliage of plants. Still fewer species are transmitted from plant to plant by insects.

It is now recognized that crop yields in most parts of the world would be significantly reduced unless steps were taken to minimize

TABLE 7.1.

Generic and common names of important plant-parasitic nematodes.

	Generic name	Common name
Endoparasitic nematodes		
Sedentary		
	Meloidogyne	Root knot
	Heterodera	Cyst
	Tylenchulus	Citrus
	Rotylenchulus	Reniform
	Anguina	Seed gall
Migratory		
	Pratylenchus	Root lesion
	Radopholus	Burrowing
	Ditylenchus	Bulb, stem, or potato rot
	Aphelenchoides	Bud, leaf
	Rhadinaphelenchus	Red ring
Semiendoparasitic, migratory nematodes		
	Helicotylenchus ⎫	
	Rotylenchus ⎬	Spiral
	Scutellonema ⎭	
	Hoplolaimus	Lance
	Paratylenchus	Pin
	Hemicycliophora	Sheath
Ectoparasitic nematodes		
	Belonolaimus	Sting
	Dolichodorus	Awl
	Trichodorus	Stubby root
	Xiphinema	Dagger
	Longidorus	Needle
	Tylenchorhynchus	Stylet

the effects of nematodes. Examples of the severity of nematode damage have been described in Florida, California, and several other regions of the world. A good example is the celery production on the irrigated sandy soils of central Florida. About 1900, the average yield was 1,000 crates per acre. Plantings were continued on the same land, and yields gradually declined. Larger amounts of fertilizer with added minor elements were used in an effort to regain original yields. Fungicides and insecticides were used more frequently. But yields continued to decline, and by 1949, the average yield was only about 300 crates per acre. In 1949 and 1950, nematodes were recognized as pathogens of celery in Florida. Research proved that by the use of nematicides, plus the other cultural practices then in use, the yield of celery could be increased to more than 900 crates per acre. Similar results have been obtained in tests with tobacco, corn, cotton, most

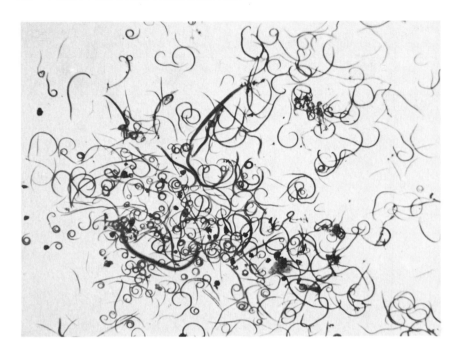

FIGURE   7.1

A photomicrograph of nematodes shortly after their removal from soil. Many are spiral nematodes, *Helicotylenchus* spp.

vegetables, and soybeans. Overall crop losses due to nematodes have recently been estimated at 6 to 12 percent in the United States (Feldmesser, 1971).

FEEDING ACTIVITIES
OF PLANT-PARASITIC NEMATODES

Every plant-parasitic nematode is equipped with a small protrusible oral stylet that enables it to penetrate plant cells and feed therein (Figure 7.2). These stylets are hollow and act much like small hypodermic needles. The hollow portions (lumens) are so small that only highly liquefied food can be taken into the body. To accomplish liquefication or partial preoral digestion, nematodes are equipped with esophageal glands (usually three in number) that produce secretions generally referred to as digestive enzymes. The largest esophageal gland, located dorsally, is believed to produce the

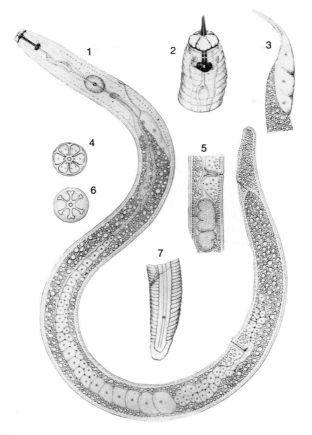

FIGURE 7.2

A typical plant parasitic nematode, *Pratylenchus brachyurus* (a root-lesion nematode). 1, female; 2, head (showing stylet); 3, basal portion of esophagus (showing three glands); 4 and 6, cross sections; 5, vulvar region (showing eggs in the uterus); 7, tail. [Courtesy of Limhuot Nong.]

biochemical materials secreted through the stylet during the processes of penetration and feeding (Figure 7.2). The glandular secretions of plant-parasitic nematodes initiate many interesting and some unique reactions by host plants. Sometimes the hosts respond by producing distinctive enlarged cells on which the parasites must feed in order to mature and reproduce (Figure 7.3). These "giant cells" (syncytia) may contain one or many nuclei, depending on the species of host or parasite involved. Other host plants react by producing many more than the normal numbers of cells surrounding the enlarged "giant cells." Consequently, an enlargement of such affected portions

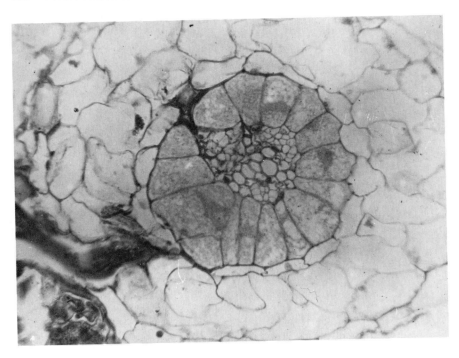

FIGURE   7.3
A ring of "giant cells" caused by feeding by the reniform nematode, *Rotylenchulus reniformis,* in a tobacco root. The anterior portion of the nematode is shown at the lower left. [Courtesy of G. C. Smart.]

of the plant occurs (Figure 7.4). These tumerous regions, usually termed galls, are composed of the enlarged "giant cells," the parasites, and the extra cells produced as a result of the parasitism.

Nematode feeding often stops cell division (mitosis) in the host tissues surrounding the feeding parasite. This usually occurs when the parasites feed at or very near the root tips (Figure 7.5). Sometimes plant cells may be killed outright by the nematodes, whereupon cavities or necrotic lesions result.

Nematode damage to host plants is ultimately biochemical in nature. Whereas some cells may be destroyed by the feeding processes, nematode secretions account for virtually all the host damage. Some adverse effects on host plants are far removed from the actual feeding sites of the parasitic nematodes. For example, plants whose roots are affected by cyst nematodes usually bear chlorotic and stunted leaves. Christie and Perry (1958) have discussed this subject in detail.

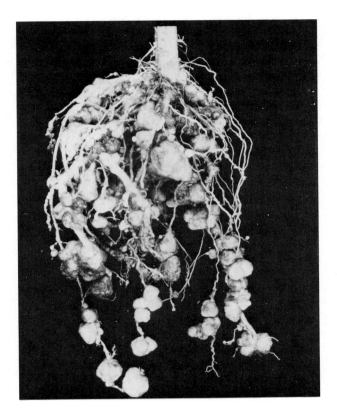

FIGURE   7.4

Tomato roots showing symptoms of root-knot, due to parasitism by *Meloidogyne incognita.*

THE LIFE CYCLES OF NEMATODES

Life cycles of most nematodes are very simple, but a few parasites of vertebrate animals have complex ones. All nematodes have an egg stage, four larval stages with four intervening molts (shedding of the outer cuticle), and an adult stage. Plant-parasitic species usually undergo the first molt (from first-stage larvae to second-stage larvae) within the egg. The second-stage larvae may overwinter or otherwise persist free in the environment (soil). The third-stage or fourth-stage larvae of some species may also be persistent. Adult nematodes also overwinter, especially within the tissues of perennial hosts. The method by which nematodes persist when no hosts are available or

FIGURE   7.5

Corn-root systems showing stubby-root symptoms as a result of feeding by the stubby-root nematode, *Trichodorus christiei.*

when environmental conditions are unfavorable depends on the species. Some, such as the burrowing nematode, *Radopholus similis* (Cobb) Thorne, can persist for only a few months without feeding and reproducing. Others, such as the cyst nematodes of the genus *Heterodera,* can persist for ten years or longer even when no host plants are available. Actually, root exudates from host plants accelerate the hatching of eggs of species of *Heterodera,* whose eggs are likely to remain dormant until exposed to the "hatching factors" contained in the root exudates from host-plant species (Calam et al., 1949).

Plant nematodes usually begin to feed on hosts during their second larval stage, which must be considered the usual infective stage. Some invade roots at this time, whereas others invade roots of their host plants as adults or during any of the larval stages. Other nematodes feed entirely from the exterior. Those which feed internally are called endoparasites, the others, ectoparasites. The females of some nematodes, usually those of endoparasites, become saccate (swollen and sac- or pouch-shaped) during their development or upon reaching the adult stage. Males retain the eelworm or vermiform shape throughout their life cycles, or become saccate and revert to the vermiform shape during the last molt.

CONTROL OF PLANT-PARASITIC NEMATODES

Students of the various disciplines involved with pest control often discuss the question of how and when the first pest-control procedures were developed. With plant-parasitic nematodes, the first control procedures were developed long before human beings even knew that there were such organisms as nematodes. One such control measure is still used in many parts of the world for banana production. Before new plantings are made, the banana corms for propagation are dug and placed in sunlight; this dries the corms and heats their outer surfaces. The corm is turned at intervals to expose all outer portions to the sun. The roots and other outer tissues, which harbor practically all the nematodes, undergo a lethal heating and drying treatment. Thus, the propagating material becomes relatively free of nematodes. Treated corms planted in land with a low concentration of the parasites produce a good crop of fruit, even though nematode concentrations may reach damaging levels at about the time of harvest.

Similarly, other control procedures such as the use of resistant cultivars and crop rotations, were developed empirically before nematodes had been recognized as plant parasites. The first cultivars resistant to nematodes survived because they had been selected for vigor and production, not necessarily for resistance to nematodes. For example, Smith (1941) claimed that some modern cultivars of cotton survived because of resistance to root-knot, even though growers and scientists alike were long unaware of the nematode. A snap-bean cultivar, Alabama #1, was collected by Isbell (1931) from a homeowner in Alabama. The variety is highly resistant to root-knot, even though it had not been knowingly selected for such resistance.

At some time prior to recorded history, people realized that continuous cropping resulted in reduced yields and therefore moved to "new" land. The world population has grown so large that new lands are no longer available; consequently, many crop-rotation schemes are now practiced throughout the world. Crop rotations, like so many other practices, have many effects other than influencing nematode populations.

The first successful methods developed by scientists for the management of plant-nematode populations were the crop-rotation procedures developed by Kühn and his associates in Germany in the 1880's. Rotations were developed for control of the sugar-beet-cyst nematode, *Heterodera schachtii* Schmidt, which caused a devastating disease called Rubenmüdigkeit or beet-root lassitude. Even today the control of cyst nematodes throughout the world is based on Kühn's recommendations. Refinements have been made, and chemical nematicides and resistant varieties are now recommended to reduce the rotation time. Since Kühn's early reports, scientists have devised many kinds of plant-nematode-control procedures, some of which are simply modern improvements of practices begun decades ago by growers.

Next I will briefly discuss contemporary control measures under the seven categories of plant-pest control listed and defined in Chapter 5. Detailed accounts of nematode-control practices appear in standard reference books and monographs (Christie, 1957; Palm, 1968; Thorne, 1961).

Exclusion

Exclusionary measures may be legally required and enforced, or they may be voluntary among individuals or groups. Legal measures in the United States for the exclusion of nematode pests began with the first quarantine laws. Several states have since begun to regulate the movement of nematode pests from outside their boundaries and within certain areas of the states. The bulb or stem nematodes of the genus *Ditylenchus* were among the first to be legally restricted, and even today much effort is made to ensure that propagating and other plant materials introduced into the United States are free of the pests belonging to that genus. These nematodes have wide host ranges, and attack plant parts above and below ground. Some species also enter anabiosis (a state of inactivity) during the fourth larval stage, and are able to survive for long periods under dry or cool conditions.

The transport of several cyst nematodes is also regulated by law. The females of these species retain most of their eggs in their body. On death, the female's outer covering, or cuticle, undergoes a change to become a highly persistent container for approximately 300 eggs, each containing a second-stage larva. These bodies (cysts) can withstand drying and freezing, and the eggs within may live for more than ten years. Eggs contained in cysts can be moved by various agents, including people, machinery, animals, seeds, birds, wind, and water. Cysts have been found on shoes, on the undercarriages of imported automobiles, and even in materials used to wrap porcelain dishes. Several important species of cyst nematodes are considered exotic to North America; if they occur in that continent at all, they are of limited distribution.

The burrowing nematode, *Radopholus similis* (Cobb) Thorne, is another nematode whose transport by people is regulated. This species causes the severe disease known as spreading decline of citrus, a disease known only in Florida. The nematode is widely distributed in tropical areas of the world, where it causes great losses in banana production and also parasitizes many other plants, including numerous ornamental species. Citrus-growing areas outside Florida prohibit the introduction of any plants infected by the burrowing nematode.

These are only a few examples of the use of exclusionary measures to prohibit the introduction of harmful nematodes into new areas. One may speculate about whether such measures are successful or not. For example, regulations to prohibit the spread of the soybean-cyst nematode were initiated soon after this pest was first detected in North Carolina during the 1950's. By the early 1970's, however, the pest had been detected in all major soybean-producing areas of the United States, and the federal quarantine measures were removed because of their ineffectiveness. The nematode probably had been distributed prior to the quarantine. A similar nematode, the golden nematode of potato, was first detected in the United States on Long Island, New York, in 1943. Quarantines were imposed, and the pest is not known in the United States outside of New York. A combination of exclusion and eradication by soil fumigants has apparently contained the potato-rot nematode to within two small seed-producing areas in Idaho and Wisconsin. In Florida, a combination of sanitation, use of chemical nematicides, and exclusionary practices have made it possible to continue shipments of ornamental plants to citrus-growing areas without spreading the burrowing nematode.

Eradication

Several measures are used as eradicative practices for the control of nematode-induced diseases of plants. The disease effects can be satisfactorily controlled without complete elimination of the parasite, although elimination has been achieved in some situations. With most annual crops, nematodes must be present in fairly large numbers (approximately 50 to 100 nematodes per 100 ml of soil) while the plants are young in order for significant yield reductions to occur. In fact, low population levels (less than 25 nematodes per 100 ml of soil) of such important nematode pathogens as the root-knot nematode have been shown to enhance plant growth, apparently by stimulating greater root production. The goal in nematode-caused plant-disease control, therefore, is to reduce the numbers at or before planting. Most nematode life cycles require three to four weeks under optimum conditions.

Consequently, when treatments are timed and applied correctly, plants can establish root systems while the numbers of nematodes are low.

The most widely used eradicative measures involve the use of chemical nematicides, but flooding, sunbaking, crop rotations, and other methods are also used. The first successful chemical nematicides used commercially were the halogenated hydrocarbon fumigants, dichloropropene-dichloropropane mixtures and ethylene dibromide, which have been widely used. Such materials are phytotoxic to most plants, and must be applied to the soil two weeks or more before planting. Fumigants containing dibromochloropropane can be safely applied to several perennial plantings, but this chemical is highly toxic to many crop plants. Generally, the fumigants are injected six to eight inches deep on 12-inch spacings, or they may be applied only along the planting row, leaving the middles untreated.

Several organophosphate and dithiocarbamate pesticides have recently proved effective in many situations for plant-nematode control (Figure 7.6). Examples include ethoprop, phenamiphos, and fensul-

FIGURE 7.6

Left and right: rows of corn growing in soil treated at planting time with an organophosphate nematicide for the control of the sting nematode, *Belonolaimus longicaudatus*. Center: rows of corn in soil that had not been chemically treated.

fothion. These can be applied as granular compounds to the soil surface, but must be mixed with the soil by cultivation equipment such as disk harrows. They are most effective if followed by rainfall or irrigation. These chemicals and fumigants have been developed because of their specific ability to kill or otherwise adversely affect nematodes so that plant parasitism is prevented. They also affect other organisms, but seldom control pests other than nematodes. Some, however, control certain soil-borne insects. For example, ethylene dibromide and mocap control wireworms. Some influence soil-borne fungi or bacteria. For example, the numbers of nitrifying bacteria may be reduced, and those of *Trichoderma* (a fungus) increased, by application of a dichloropropene fumigant. Enhancing the growth of *Trichoderma* is beneficial, for that saprophyte is antagonistic to soil-borne plant parasites.

Much research has been directed toward finding chemicals that are effective against multiple pest populations. For example, methyl bromide and chloropicrin are soil fumigants used to control a broad spectrum of pest species. Although especially effective against soil nematodes, they also give excellent control of weeds, insects, fungi, and bacteria. Some mixtures of nematicides and fungicides have been used to control nematodes and fungi simultaneously. Examples include control of root-knot nematodes and *Fusarium* wilt of tomatoes, and control of nematodes and the black-shank disease of tobacco.

Research and experience have proved that several cultural practices are effective in keeping nematode populations well below their economic thresholds. One of the most widely used practices is crop rotation (Figure 7.7). Earlier mention was made of the research in Germany and elsewhere to provide crop-rotation schemes for controlling the sugar-beet-cyst nematode. Cyst nematodes (*Heterodera* spp.) usually have more limited host ranges than do other nematode pests; consequently, crop-rotation schemes are easy to devise and convenient to use. Cyst nematodes, however, can persist as eggs within the cysts for several years under some conditions, and long rotation periods are required. Other nematodes, such as *Meloidogyne* spp., have very wide host ranges and will reproduce on many cultivated crops, weeds, or wild host plants. Thus, successful rotation schemes must be carefully planned for specific geographical areas. Also, the land involved must be kept free of weed hosts at all times.

Some plants, such as asparagus, marigold, and pangolagrass, produce toxins that adversely affect some stages of certain nematodes. Pangolagrass is especially effective in reducing the numbers of root-knot nematodes, and is used as a pasture crop in rotation with row crops. Young roots produce materials that increase hatching rates of

FIGURE   7.7

A field of peanuts in southern Alabama. Peanuts were planted for the first time in that part of the field shown in the upper left of the photograph. Peanuts had been grown every year for five years in the remainder of the field.

root-knot nematode eggs. Older roots were found to produce materials toxic to the root-knot larvae (Winchester and Hayslip, 1960). Little else is known about these interesting phenomena. Other plants, such as *Crotalaria spectabilis*, effectively reduce populations of most nematodes when used as rotation crops. For example, larvae of root-knot nematodes invade roots of *C. spectabilis* and establish a parasitic relationsip, but they do not mature and reproduce. Thus, this plant is an excellent trap crop (Steiner, 1941).

There is much variation in host ranges between species of nematodes. For example, the sting nematode, *Belonolaimus longicaudatus* Rau, is a major pest of peanut in sandy soils of Virginia and North Carolina. But this species does not attack peanut in Florida and Georgia, where peanut can be rotated successfully with susceptible hosts such as corn and beans. Conceivably, different strains of *B. longicaudatus*, each having a different host range from the other, occur in different regions of the United States.

Nematodes are fragile organisms and are very susceptible to sudden, drastic changes in their environment. Rapid drying and exposure to sunlight have often been used for their control. Continuous flooding of land for as long as four months has been effective in reducing the numbers of most plant nematodes (Overman, 1964). Alternation of flooding and fallowing has sometimes been more effective than either practice alone (Rhoades, 1964).

Therapy

Several means for reducing numbers of plant nematodes in and on their hosts have been developed. Chemicals, hot water, sunbaking, and mulches are routinely used. Two chemical nematicides, fensulfothion and phenamiphos, are relatively nonphytotoxic and may be safely used for therapeutic treatment of certain plants. Environmental factors and the sensitivity of certain other plants must be considered before the use of any chemical for application to established plants or propagating materials.

One of the most successful therapeutic treatments for nematode control is to use one of the several nematicides available for control of the nematode pests of turfgrasses. It is estimated that nematicides are applied annually to greens and fairways of about 80 percent of the golf courses in Florida. Dibromochloropropane and phenamiphos are commonly used to control such nematodes as sting, lance, ring, and root-knot. Nematicides applied to ornamental and other plants improve their quality. Nematicidal therapy has sometimes allowed the lifting of quarantine restrictions, so that treated plants could be moved from restricted regions.

The use of green-manure mulches around perennial plants may reduce the rates of increase in populations of plant-parasitic nematodes. Mulches applied to orange trees in Florida, for example, have resulted in reduced populations of *Tylenchulus semipenetrans*, with corresponding improvement in tree growth and fruit production. Gardenias, figs, and peaches infected by root-knot have shown similar responses to mulches. It used to be thought that the organic mulches increased the populations of biological control agents; however, recent research (Hunt et al., 1973) showed that toxic decomposition products of organic matter were responsible for reducing the populations of nematodes.

Plant Resistance

No effort will be made here to separate resistance to plant nematodes into horizontal and vertical resistance, because genetic inheritances have not usually been established. Apparently, resistance to nematodes is usually vertical, because different biotypes "break" the resistance. For example, resistance to the soybean-cyst nematode in soybeans and to root-knot nematodes in certain peach stocks is effective only against certain strains (pathological races) of the nematodes.

A few crop varieties that are resistant to certain nematode pathogens have already been developed. Examples include soybeans resistant to root-knot, alfalfa resistant to stem nematodes, and cotton resistant to root-knot. Plant breeding for nematode resistance offers promise for the future. To be successful, however, the plant breeder must consider that several nematode pest species occur in all agricultural soils. Also, the pathogenic variability within individual species must be discovered before crop varieties that are resistant to predominant strains can be developed (Figure 7.8).

Protection

Protective measures as defined in Chapter 5 have not been developed for nematode control, but some of the measures discussed under therapy may also afford protection. For example, nematicides contain-

FIGURE   7.8

Citrus seedlings grown in soils infested by four different populations of the sting nematode, *Belonolaimus longicaudatus*. Only one population *(F)* stunted the growth of seedlings.

ing dibromochloropropane not only kill nematodes, but also appear to prevent the nematodes from finding host roots. Some systemic nematicides have limited use and may both kill (therapeutically) nematodes that are already established as parasites and protect treated plants from subsequent attacks.

Avoidance

Early or late plantings are sometimes made to avoid plant nematodes. Potatoes often escape serious injury from root-knot when planted early in the spring, while soil temperatures are still low (20°C). Tubers are believed to be attacked by secondary larvae produced on roots. The life cycle of these nematodes is greatly lengthened by reduced soil temperatures. This reduces the number of parasites that can attack the tubers as they form. In Florida, *Aphelenchoides besseyi* can be very damaging to strawberries from June to September. Later plantings, normal for Florida, completely escape attacks by this pest because temperatures, especially at night, are too low for the nematode (Christie, 1942).

LITERATURE CITED

Calam, C. T., H. Raistrick, and A. R. Todd. 1949. "The potato eelworm hatching factor, I: The preparation of concentrates of the hatching factor and a method of bioassay." *Biochem. J.*, 45:513–519.

Chitwood, B. B., and M. B. Chitwood. 1950. *An Introduction to Nematology*, rev. ed. Baltimore: Monumental Printing.

Christie, J.R. 1942. "A description of *Aphelenchoides besseyi*, n. sp., the summer dwarf nematode of strawberries, with comments on the identity of *Aphelenchoides subtenuis* (Cobb, 1926) and *Aphelenchoides hodsoni* (Goodey, 1935)." *Proc. Helminth. Soc. Wash.*, 9:82–84.

_____. 1959. *Plant Nematodes, Their Bionomics and Control*. Gainesville: Fla. Agr. Expt. Sta.

_____, and V. G. Perry. 1958. "Mechanism of nematode injury to plants." In *Plant Pathology: Problems and Progress, 1908–1959*. Madison: Univ. of Wisconsin Press.

Feldmesser, J., chairman. 1971. *Estimated Crop Losses from Plant-Parasitic Nematodes in the United States*. Soc. of Nematologists. Special Publication No. 1. Beltsville, MD.

Hunt, P. G., G. C. Smart, Jr., and C. F. Eno. 1973. "Sting nematode, *Belonolaimus longicaudatus*, immotility induced by extracts of composted municipal refuse." *J. Nematol.*, 5:60–63.

Isbell, C. L. 1931. "Nematode-resistance studies with pole snap beans." *J. Heredity*, 22:191–198.

Overman, A. J. 1964. "The effect of temperature and flooding on nematode survival in fallow soil." *Proc. Soil and Crop Sci. Soc. Fla.*, 24:141–149.

Palm, C. E., chairman. 1968. *Principles of Plant and Animal Pest Control, Vol. 4: Control of Plant-Parasitic Nematodes.* Publication 1696. Washington, DC: National Academy of Sciences.

Rhoades, H. L. 1964. "Effect of fallowing and flooding on root-knot in peat soil." *Plant Dis. Rep.*, 48:303–306.

Smith, A. L. 1941. "The reaction of cotton varieties to Fusarium wilt and root-knot nematodes." *Phytopathology*, 31:1099–1107.

Steiner, G. 1941. "Nematodes parasitic on and associated with roots of marigolds (Tagetes hybrids)." *Proc. Biol. Soc. Washington*, 54:31–34.

Thorne, G. 1961. *Principles of Nematology.* New York: McGraw-Hill.

Winchester, J.A., and N. C. Hayslip. 1960. "The effect of land-management practice on the root-knot nematode, *Meloidogyne incognita acrita*, in south Florida." *Proc. Fla. Hort. Soc.*, 73:100–104.

8

Entomology and the Control of Arthropod Pests of Plants

Sidney L. Poe

Entomology is the division of zoology that deals with insects and their classification, structure, functions, distribution, population dynamics, potential usefulness or destructiveness, and population control. Although entomology is restricted by definition to mean the study of insects, the realm of applied entomology for pest-control purposes has been extended to include almost all invertebrate pests: snails, slugs, spiders, spider mites, and scorpions. Applied entomology is a descriptive name used to identify and characterize activities aimed at solving problems caused by invertebrate species that have become pests because of their harmful interactions or competition with crop plants. How entomologists deal with potentially harmful pests is the science and art of pest control and is the subject of this chapter.

The application of entomological techniques for the control of pests should be based on logical assumptions or data that describe the species one hopes to influence. Population suppression may be fortuitous, but thorough knowledge of individual pest species—their life histories and cycles in nature, their host plants and animal relations—greatly increases the probability of successful pest control by applied techniques. Much research is often needed to understand the life cycles of pestiferous species, and only when such data are available can a rational strategy be devised and appropriate tactics used to control them. Based on the fact that female screwworm flies mate only once for life, a program for the mass release of sterile male screwworm flies was conceived and funded. This ultimately became one of the most successful insect-control programs in history (Knipling, 1955). Because sustained control is based on detailed knowledge of pest species, recognition of the basic biological factors that influence insect populations, their development, and their role in nature is essential.

INSECTS IN THE ECOSYSTEM

Invertebrate animals are components of an intricate network through which energy flows from green plants (producers) to primary, secondary, and tertiary consumers of energy. Insects and mites may be found at every level of energy transfer, including the saprophytic and decomposer levels in the food chain. Insects can be phytophagous and feed directly on plants themselves, as the gypsy moth does on deciduous trees, or they can prey on other plant-feeding insects. Others may prey on the predator of the primary phytophagous species. Many insects consume organic material, either plant or animal remains, and thus contribute to the decomposition of organic matter and the formation of humus. Insects also attack saprophytic fungi and other organisms that assist in the breakdown and decay or organic materials.

Thus, by such an ecological system, solar energy is trapped by green plants and converted into the chemical energy used for building plant materials. In turn, the consumers may be eaten or destroyed, and energy is returned to the soil for the support of plants, which then are able to trap more solar energy. The cycle established by this energy flow is repeated again and again, and is a segment of a complex ecosystem. Each ecosystem is maintained in a stable condition by the presence and function of many interdependent species, and, although populations of one species or another may be dominant, there are enough different species to maintain diversity, which leads to a certain degree of uniformity and stability. Populations of all organisms in such an energy-flow network have appropriate checks and balances that stabilize their numbers.

An imbalance occurs when one or more species becomes extremely scarce or extremely numerous. Such an imbalance may result from cataclysmic events that alter a resource, that remove a check or balance, or both. All living species of plants and animals compete for the resources in natural ecosystems. Human beings are no exception and must battle the elements of nature and other species for any life-sustaining commodities they obtain. It becomes paramount to understand the intricate balance of the driving ecosystem forces in nature, which operate on insects as well as on other fauna and flora to provide an "edge" for human beings in their bid for natural resources. Advantage is based on thorough and complete understanding of the adversary.

Of the myriad species of plants and animals found in natural ecosystems, about 80 percent are arthropods. Between 800,000 and one million species of insects have been catalogued and described (Table 8.1). Fortunately, only a few of this number pose serious threats to human beings. Many species are beneficial; other species are innocuous and do not directly benefit or harm humans. Many beneficial insects consume, by predation or parasitism, the arthropods that compete with humans. Many species may be harmful to human beings or to crops only during one phase of their life histories or development. Seventeen-year cicadas destroy twigs and branches with their swordlike ovipositors, but appear only every 17 years. Species of the Lepi-

TABLE 8.1.

Major orders of the Insecta.[a]

Order	Example	Metamorphosis	Mouthtype	Wings	Number of species
Protura	Protura	Ametabolous	Suctorial	None	118
Thysanura	Silverfish	Ametabolous	Chewing	None	370
Collembola	Springtail	Ametabolous	Suctorial	None	2,000
Ephemeroptera	Mayfly	Incomplete	Vestigial	2 pr	2,000
Odonata	Dragonfly	Incomplete	Chewing	2 pr	4,870
Orthoptera	Grasshopper	Gradual	Chewing	2 pr	22,500
Isoptera	Termite	Gradual	Chewing	2 pr	1,100
Dermaptera	Earwig	Gradual	Chewing	2 pr	1,100
Plecoptera	Stonefly	Gradual	Chewing	2 pr	1,550
Psocoptera	Book louse	Gradual	Chewing	2 pr	1,100
Mallophaga	Chewing louse	Gradual	Chewing	None	2,675
Anoplura	Sucking louse	Gradual	Sucking	None	250
Thysanoptera	Thrips	Gradual	Rasping	2 pr	4,500
Hemiptera	Stinkbug	Gradual	Sucking	2 pr	23,000
Homoptera	Aphid	Gradual	Sucking	2 pr	32,000
Coleoptera	Beetle	Complete	Chewing	2 pr	290,000
Mecoptera	Scorpion fly	Complete	Chewing	2 pr	400
Neuroptera	Lacewing	Complete	Chewing	2 pr	4,670
Trichoptera	Caddis fly	Complete	Siphoning	2 pr	4,450
Lepidoptera	Butterfly	Complete	Sponging	2 pr	112,000
Diptera	Housefly	Complete	Biting, sucking	1 pr	86,000
Siphonaptera	Flea	Complete	Sucking	None	1,100
Hymenoptera	Bee	Complete	Chewing	2 pr	105,000
Miscellaneous	—	—	—	—	900

[a]*Source:* Borror and Delong (1971).

doptera (moths and butterflies) are destructive as caterpillars; but eggs and pupae are not offensive, and adult forms are even cherished.

Wherever competition between insects or mites and human beings has become intense enough, people have devised tactics which, when applied to the commodity or to the pest, have reduced the imposed threat. Applied entomology and crop-protection methods have arisen out of the need for better insect control. Methods for control have been developed for use in homes, on pets, on livestock, on farms, in barns; in fact, in any situation where insects compete with people for a natural resource.

THE NATURE OF INSECTS

Insects are classified as arthropods; they have paired, jointed legs, and possess a segmented body enclosed by a rigid sclerotized (hard) exoskeleton, made flexible by non-sclerotized areas between segments. The Insecta (Hexapoda) is the only class of invertebrates whose members possess wing structures that enable many species to fly during parts of their life cycles. Their bodies are readily divisible into three distinct regions: head, thorax, and abdomen.

The anterior or head region contains the primitive brain and bears several external sensory organs: the compound eyes, and usually simple eyes or ocelli; the antennae, wtih their complement of mechano-sensory and chemosensory hairs, spines, and projections; and the paired mouth parts. Each compound eye is composed of numerous facets or units, which function together to provide a mosaic vision. Simple eyes are sensitive to light intensities, but do not form images (Chapman, 1971).

The mouthparts are variously modified in the several orders of insects (Table 8.1) for ingesting and handling food. The most unspecialized, basic form is the chewing mouth, which consists of paired structural appendages adequate to, shred, handle, and grind food (Figure 8.1). The jaws (mandibles) move from side to side, and, with the aid of maxillae, tear and masticate food. The chewing-mouth type is well-illustrated in common species of beetles (Figure 8.2), grasshoppers, and caterpillars (Figure 8.3). Modifications of this basic mouth type are found in other insect orders.

Aphids, scales, leafhoppers, cicadas, stink bugs, and mosquitoes have piercing-sucking mouth types (Figure 8.4). The same basic appendages as are functional in chewing-type mouth parts are present,

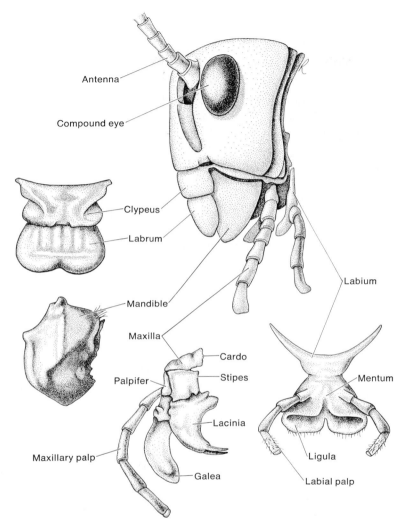

FIGURE   8.1
Diagram of insect mouthparts used for chewing.

but they are highly modified and sheathed as stylets in a bundle, forming two channels for the passage of saliva and food materials. The stylets are drawn out into a slender beak used for piercing host tissues (Figure 8.5). When a suitable depth is reached, a sucking action is initiated, and liquid food is removed from the host. In aphids and

FIGURE 8.2

Mexican bean beetles and bean leaves injured by their feeding. [Courtesy of the Florida Department of Agriculture and Consumer Services.]

similar species, the pressure of fluid in plant vascular elements may be great enough that no sucking by the insects is required to obtain nourishment.

Many pathogens are transmitted by insects with piercing-sucking mouth types. Probing action of the slender stylets in piercing the tissues provides entry for pathogens such as viruses, MLO's, or rickettsiae; so the threat posed by these insects may be both primary and secondary. Foremost among insect-transmitted pathogens are those that cause human diseases like malaria and yellow fever, and plant viral diseases like mosaics and yellows. An intimate association may exist between the insect, its salivary mechanism, and the pathogen. Knowledge of the association between pathogen transmission and the insect vector has led to the popular concept of disease control by control of the vector, which has been successful in malarial mosquito-control programs.

Thrips have specialized mouth types in which one mandible is reduced, and the other developed to rasp tender tissues. Houseflies have a sponging-lapping type, which functions to sponge up liquid or digested food. Butterflies and moths extract nectar and fluids from flowers and other plant parts through a slender, coiled tube. Larvae of some moths "mine" the mesophyll tissues of leaves (Figure 8.6).

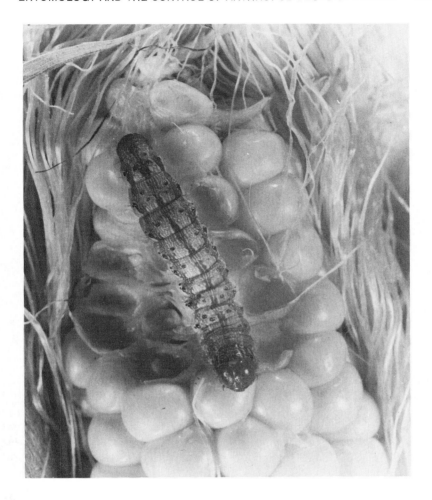

FIGURE   8.3
The corn earworm and the injury its feeding causes in developing kernels of corn.
[Courtesy of the Florida Department of Agriculture and Consumer Services.]

The mouthparts of mites are unlike those of the insects: they are derived in an altogether different embryological sequence. However, plants are injured when mites insert their needle-like stylets into tissues and individual cells to remove fluids. Such action, of course, destroys the cell, and the feeding of numerous mites soon results in a fleck-and-brown pattern as the cells die. A mite-damaged leaf, bud, or scale may be aborted by the host plant, and flowers or fruits may be

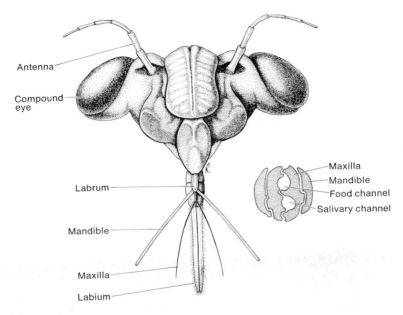

FIGURE   8.4
Diagram of the piercing-sucking mouthparts of some insects.

left deformed. The principal families of plant mites are listed in Table 8.2.

The middle region of insects, the three-segmented thorax, bears three pairs of legs and, in winged forms, two pairs of wings or their modifications. Insect legs have been variously adapted to perform different functions: fossorial for digging (molecrickets), flat and fringed for swimming (aquatic insects), enlarged for leaping (grasshoppers, crickets). Wings are not present in some insects (fleas, lice); some have only two wings (flies); others have four; still others possess wings only for a short time for special mating or dispersal activities (termites, ants, aphids).

The posterior region of the insect body is the abdomen, consisting of twelve segments, usually variously modified and coalesced. Reproductive appendages are borne terminally on the abdomen and are also variously adapted to deposit eggs or, as with the stingers of ants and wasps, to paralyze prey or to defend the nest of a social group.

In the digestive system of insects, food moves from the mouth into an alimentary canal, which consists of fore, mid, and hind regions. Various enzymatic processes occur, and digested food material is absorbed directly into the blood from the midgut region, as waste excreta

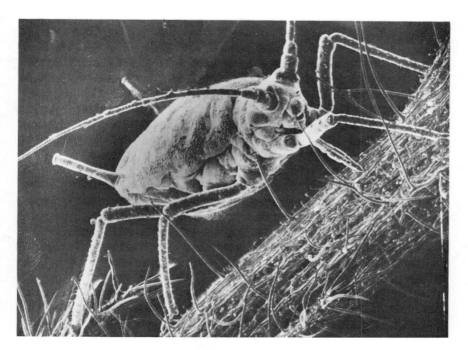

FIGURE 8.5

An aphid feeding on a plant stem. [Scanning-electron microphotograph courtesy of USDA.]

FIGURE 8.6

Serpentine trails in the foliage of chrysanthemum caused by the feeding of larvae of leafminers.

TABLE 8.2.
Acarine families containing important plant-feeding species of mites.

Family	Common names	Host plants
Tetranychidae	Spider, red spider mites	Fruit, vegetables, berries
Tenuipalpidae	False spider, flat mites	Fruit, ornamentals
Tuckerellidae	Tuckerellids	Ornamentals, citrus
Tarsonemidae	Fungus, flower mites	Fruit, ornamentals
Pyemotidae	Hay-itch mites, pyemotids	Ornamentals, cereals
Pentheloidae	Winter-grain mites	Cereals, legumes
Tydeidae	Honeydew mites	Citrus, fruit
Eriophyidae	Bud, gall, blister, rust mites	Fruit, cereals, ornamentals
Acaridae	Bulb, root, soil mites	Bulb, fleshy root crops

are transported posteriorly to be voided. The digestive system and alimentary canal are highly modified in various groups of insects, but adaptations and the complement of enzymes used to break down food generally reflect the nature and kinds of foods eaten.

The circulatory system of insects is open, consisting of a pulsating heart located dorsally in the abdomen and an aorta extending forward through the thorax into the head. The blood then moves sluggishly toward the rear of the body cavity, where organs and tissues are bathed. Blood does not transport respiratory gases, but does transport nitrogenous wastes, acts as hydraulic-pressure systems, and distributes digested food. A primitive protective function is also served by clotting and the closing of wounds by action of the elements in the blood. Malphigian tubules extract nitrogenous wastes from the blood and void it through the alimentary canal.

The nervous system is not very sophisticated; it consists of a single brain and paired ventral nerve cords, with ganglia (nerve centers) in each region or segment of the body. Each segment with its ganglion acts somewhat independently, and neural branches arising from the ganglion pass to the functional parts of each segment. Neuromuscular junctions function much like those of vertebrate nervous systems.

Respiration, or, more appropriately, "breathing," is accomplished by a system of hollow trachea and ramifying tracheoles that reach the interior of the insect body from lateral tracheal trunks. Air enters the tracheal trunk through paired openings, the spiracles, located in segments of the thorax and abdomen. Oxygen diffuses into the tracheal system or is drawn through it to the tissues; waste gasses are removed from the tissues along the same pathways.

The details of insect systems have been described in standard references (Chapman, 1971; Metcalf et al., 1962).

GROWTH AND DEVELOPMENT OF INSECTS

Metamorphosis

Insects undergo changes in form as they increase in size and maturity. These changes, of four basic types, are described as metamorphosis. A few primitive orders (Table 8.1) have ametabolous metamorphosis; that is, they develop with essentially no structural change. Their growth is an increase in size and maturation of reproductive organs. Adults and juveniles live and feed together.

Orders of insects with aquatic nymphs (dragonflies, mayflies) show incomplete development. The young nymphs (naiads) are vastly different from the adults; they live in a different habitat and consume different foods.

Gradual development occurs in those insects with three distinct forms: egg, nymph, adult. The nymphs resemble adults, but lack wings and mature gonads. Their life progresses through several juvenile stages of increase in body size until maturity. Nymphs live and feed with the adults, and such insects as nymphal grasshoppers, crickets, earwigs, or aphids can be easily identified because they resemble the adults. Although they undergo a similar kind of metamorphosis, the difference between nymphal and mature whiteflies, cicadas, or thrips may be striking.

Insects with complete metamorphosis are termed holometabolous insects. Their metamorphosis consists of development in four stages: egg, larva, pupa, and adult, each usually much different from the other. The larval stage is a feeding stage, and is remarkably different from the adult in appearance, usually in habitat, and in food consumed. Whereas nymph and adult of the gradual-developing insect may compete for space, shelter, and food, the holometabolous insect adult, except for a few species (primarily beetles), does not compete for food, space, or shelter with its young.

Fully fed larvae usually form a cell or cocoon, and the pupal stage (in which the larva is transformed into an adult) is initiated. Some characteristic names have been given to the larvae of the various holometabolous insects. Maggots are the larvae of flies; grubs are the larvae of beetles; and caterpillars are the larvae of moths and butterflies. Certain more specific names, such as botts, warbles, wigglers, cutworms, and trashbearers, may also be applied to larvae of the holometabolous insects.

Larvae are generally restricted to the area near the place where eggs were laid or to the diet available when they hatch from the egg. Later stages may exhibit migratory behavior. Therefore, the adult insect usually selects the host site, plant or animal, for oviposition, and most possess some means of choosing an adequate host for the young, even though the adult female might require an altogether different host. This behavioral instinct is remarkable in those species with very narrow larval-host ranges.

Growth

Insects reproduce sexually or parthenogenetically, are oviparous, or ovoviviparous, and may oviposit singly or in masses. There may be one, two, or many overlapping generations during the time when weather is favorable and resources are plentiful. As an insect grows inside its inflexible exoskeleton, it must periodically shed the old skeleton and form a new one; that is, it must molt. This molting process is under neural and hormonal control. Each molt results in an increase in size and is usually accomplished within a few hours. The insect may be rendered inactive, however, during the operation of the chemical mechanisms that dissolve the various layers of the old exoskeleton and form a new one. The molt is followed by a time during which the new cuticle is tanned and hardened. The molting process is repeated a definite number of times by growing insects until the adult stage has been reached.

Life Cycles

The life history of a generation of insects is often referred to as the life cycle, since the events involved, whether perennial, annual, or weekly, represent a cycle of growth and development. From the egg or first-stage larva or nymph, individuals progress through an orderly sequence of development to maturity. Mating and egg-laying or birth follow, and the cycle is completed.

Although the technical terminology used to describe mite life histories and developmental stages differs from that used for insects, the sequence of events is similar. Most mites develop as do insects, with gradual metamorphosis, in that egg, nymph, and adult stages are generally distinguishable within a single colony.

POPULATION DYNAMICS OF INSECTS

Although relatively few species of insects and mites interfere directly with production of crop plants, those that do frequently occur in large interbreeding and interacting populations. The presence of large numbers of individuals is a direct result of reduction of stress or re-tarding factors, which restrict population size, or of limiting factors, which slow development. Studies of the dynamic factors that influ-ence population numbers and of the genetics of species constitute the subject of population biology.

Insects, like most organisms, produce an abundance of eggs, most of which fail to reach reproductive age because of the limitations im-posed by natural forces operating within the ecosystem. Natural limitations temper extremes and maintain balance in population numbers. Overabundance of a population results in a shortage of food, an increase of natural enemies, or both. Either result will then reduce the population to a size in line with the energy resources available.

Changes in an insect population become evident when the numbers in progeny generations are greater than those in parental generations or when the converse occurs. The change in numbers from generation to generation is far from simple, but may be estimated according to certain well-defined equations. If the number of insects or mites is obtained periodically by some sampling procedure, a population curve can be drawn as discussed in Chapter 4. Such a curve might represent any stage of development (egg, nymph, larva, adult), or a composite of all stages, and can be used to show the variation in numbers in successive generations in time. Each line space or incre-ment of the curve is defined by the equation

$$N_t = N_0\, e^{rt}$$

where t is an interval of time, N_t is the number of individuals after t, N_0 is the original number of individuals, r is the intrinsic rate of popula-tion increase, and e is a constant, the base of natural logarithms or 2.71828.

This general equation, as we saw in Chapter 4, describes the change in time of the population of individuals. The rate of change, r, is the difference between number of births (natality) and deaths (mortality), or $b - d$. Such an equation can be elaborated to reflect the loss due to emigration (E) and the increase due to immigration (I), so that the total population change is given by the equation

$$N_t = N_0 e^{rt} - E_t + I_t$$

It is easy to see that a change in either component used to determine r, birth or death rate, as well as by loss due to emigration or gain due to immigration, will alter population numbers.

Birth rates of insects and mites are influenced by many factors; among the more important are fecundity, fertility, and sex ratio (Clark et al., 1967). Fecundity is the ability of females to produce eggs. The potential number of eggs produced by a species may be determined by the nature of the insect life cycle or by adequacy of its diet as an immature or adult animal. Fertility is the actual number of progeny produced when the potential is partially realized. Not all eggs are fertilized, and not all fertilized eggs hatch; so these factors affect population numbers and dynamics. Sex ratios reflect the numbers of females to the numbers of males in the population. Generally, the larger the proportion of females to males, the more rapid the population increase.

Death rates are also affected by many factors, which, in turn, alter population numbers and are thus critical to any study of population dynamics. Important causes of death may be age, low vitality, natural enemies (pathogens, predators, and parasites), food shortages, and overcrowding.

Population numbers can be affected by the spread of individuals into new areas, as well as by restrictions imposed on the population by the stresses and limitations of the environment of the ecosystem. In general, the death rate increases as the population grows, because of reduction of available food and space, and increase in epidemics, natural enemies, and other retarding and limiting factors. For every environment or ecosystem, there is a maximum value for the population. Symbolized by the letter K, this value is in actuality the carrying capacity of the environment for the population of that species (Chapter 4). The carrying capacity represents the upper limit of population stability and equilibrium.

Populations react in one of two basic patterns, as "r strategists" or as "K strategists" (Wilson and Bossert, 1971). An r strategist has a rapid life cycle and a high r value, and is adapted for life in an ephemeral habitat. Such a species can rapidly locate a suitable habitat, use the resources of the habitat and reproduce, then migrate in search of new habitats. Most of the serious insect pests of annual agricultural crops are r strategists.

The K strategist, however, is a species that thrives in a stable, diverse habitat within a complex of populations. Such species are common to climax communities and maintain their populations at levels near the carrying capacities of their environments. A high r value is of less advantage than the ability to compete for a resource in a stable environment. K strategists are sometimes present in perennial crops such as apples, citrus, or forests.

By characterizing population dynamics, plant-protection entomologists are able to recognize the factors that limit and retard pest populations, and to devise and apply methods of control.

THE NEED FOR INSECT-PEST CONTROL

Invertebrate animals choose host plants or animals for the most basic of reasons: to obtain food or protection (overwintering, oviposition), or to complete the life cycle. In so doing, these animals usually cause some injury to their hosts. The activities of human beings in modern agriculture have created conditions most conducive to widespread pest attacks on our crop plants. The concentrated large acreages of a single crop in successive years (for example, cotton and corn) have led to a general area-wide population of the pests of that crop. Genetic improvement of crop plants that grow and produce high yields rapidly for our use has also been exploited by the pests.

Excessive populations of insects may result when a potential pest species is introduced into a new area where it has no natural enemies. Indeed, many of the serious pests of North American agricuture have been introduced from another continent. For example, the alfalfa weevil, *Hypera postica* (Gyllenhal), reported from Utah in 1904, probably entered the United States from Europe. By 1970 it was present in 48 states, where it severely injures alfalfa and several other legumes (Pfadt, 1971).

Populations develop when environmental components favor rapid reproduction and where food is abundant and the weather favorable. The environment may be improved for the insect by poor cultural practices or unwise use of insecticides, which often affect natural enemies more than they affect the pest itself.

PRINCIPLES OF INSECT AND MITE CONTROL

Population controls for insect and mite pests have historically been classified by method as either natural or applied, but not necessarily according to the principles upon which each rests.

Natural Control

Natural control methods are those whose existence and operation are largely independent of human activities. They include topographic features of the landscape that serve as barriers to insect dispersion and thus provide population control. Large bodies of water, mountain ranges, and soil type and other characteristics often limit the range of pest species; these are natural checks to insect invasion or spread. These features are easily recognized as exclusionary, because they exclude pest populations from entry and thereby prevent establishment in new areas.

The effects of climate on insects may also be classed as exclusionary. Weather extremes (aridity, heat, or cold) sometimes extend beyond the ranges of insect adaptability; so populations never become established in the new territory. Such effects may also be eradicative, when they are intense enough to annihilate a population. Floods, heavy rains, hurricanes, and winds have such adverse effects on many insect populations that they may need several months or years to reach their former size. These natural forces are described as density-independent factors, because their effect is neither lessened nor intensified by the number of individuals in the population.

The Mediterranean fruit fly *(Ceratitus capitata)* is a serious pest in areas of the world that lie within the January isotherm of 10°C; it cannot become established in regions where mean temperatures fall below 10°C for three consecutive months (Martin, 1973). The Mexican bean beetle, *Epilachna varivesta* Mulsant, on the other hand, is not restricted, but has extended its range over a wide area represented by climatic extremes.

Various natural enemies of pest species exist in almost every area of the world, and these, too, exert a degree of control on the population. Since both host-specific and nonspecific natural enemies may be present, and because the efficiency of many natural enemies is reduced by low prey densities, complete destruction of the pest species is almost never accomplished. Instead, prey and natural-enemy populations oscillate at low densities under natural conditions. Birds, mammals, reptiles, spiders, insects, and mites take a heavy toll on insect-pest populations, and their effects are eradicative. Since the number of pests killed by natural enemies is proportional to the population level, their effect is density-dependent.

Other enemies of insect and mite populations include pathogenic fungi, bacteria, and other microorganisms that cause disease and kill individuals under natural conditions. Pathogens are largely density-dependent, being more efficient at high population densities. Effects of native predators, parasites, and pathogens should not be considered to result from applied measures; they operate without human assistance, often, in spite of our activities.

The long association of invertebrate animals with native flora has resulted in an orderly or stable transfer or energy among these individuals. Of the million or so species of insects known to science, many attack plants, but only a few species attack any one species of plant. Therefore, most plants are resistant to the ravages of most species of insects. An exception occurs when an omnivorous species such as the locust develops large populations and, in migration, consumes vegetation of any type that is encountered. The resistance of native flora to most pest species is both vertical, which results in little or no attack on the plant by the given pest, and horizontal, which, because of various intrinsic physical and nutritional qualities, suppresses population growth or individual development of the pest species.

Natural population controls, such as climate, weather and topographic barriers, operate independently of human activities in agriculture; they are physical phenomena that exert physical pressures against pest species. The relationship of natural enemies, either vertebrate or invertebrate, to the pest is biological, however, and populations are influenced by biological phenomena. The size of predator populations in a given year, the level of pathogen inoculum, or the presence of alternative food sources may be reflected in the size of the pest population. Furthermore, because biological interrelationships are established between the pest population and its natural enemies, human activities can become disruptive, and destroy (or restore) any balance established under natural conditions. The modern farm practice of growing plants as monocultures with little plant diversity has led to the concentration of pests, often without their natural enemies. Natural enemies may be eliminated from monocultured systems by the use of pesticides, and entomophthorous fungi may be suppressd by the use of toxic chemicals. But selective timing of insecticidal and fungicidal applications, and the judicious choice of toxicant concentrations and types, have done much to conserve the balance established by natural control.

Insects have become serious pests of new crops bred or developed for high yields. High-yielding plants that have been bred to grow rapidly and produce quickly, and that have, in the process, lost most of the genetic heterogeneity and variation, have become important sources of food for insects as well as for humans. Species present on wild plants will rapidly invade cultivated fields and adapt to well-cultured, fertilized, and nutritious hosts when the opportunity arises. The Colorado potato beetle, *Leptinotarsa decemlineata* Say, once an innocuous beetle on wild solanaceous plants in the western United States, adapted to the introduced potato, spread rapidly to the east coast, and became a devastating pest wherever potatoes are grown. This species may also be a pest of other solanaceous crop plants, notably tomato and eggplant (Metalf et al., 1962).

Development of methods for pest-population control that are based on varied scientific principles was largely abandoned when synthetic organic insecticides were introduced. These products, sprayed on crop plants or on pests, prevented population development and high yield losses. Reliance on this single method of insect-population control led applied entomology from a mere subsistence phase into exploitation and crisis phases, and, for certain crops, into a disaster phase (Smith, 1969; Metcalf and Luckmann, 1975). These various phases have been outlined to illustrate the "progress" of pest control in agriculture. But as insects become resistant to insecticides, and as the effects of insecticides on the environment and other plants and animals become intolerable, each phase becomes more extreme, until complete disaster is reached. Fortunately, all these phases have been observed in only a few crops. Such conditions have resulted primarily from imbalance in the use of the various methods and principles of population control. It is certain that future agricultural production must be accomplished under a managed system of balanced efforts to control pest populations, and not by total reliance on a single method for population control. Such a managed system will require analysis of the principles of pest control to be used in each situation in order to choose the most advantageous means—reliable means that are in tune with economics and with the safety of people and animals.

Applied Methods

Pest-control measures are applied to bring about a detrimental effect on pest populations for a given period of time in a specific location. Methods used primarily for their cultural value at the beginning of the growing season may also adversely affect pest populations and be of secondary benefit. Tilling, for example, is essential to provide aeration and a loose medium for growth of crop-plant roots, but tillage also exposes soil-borne pests to the rigors of weather and to predators. The choice of crop-plant cultivar may be based on high yield potential, marketability, or horticultural quality; any resistance to insects is a bonus. Pest-control principles can be divided into two major categories: methods of reducing the population and methods of slowing the rate of population increase (Chapter 5).

Exclusion

The principle of exclusion may take the form of legislative efforts to prevent the entry of pests into a crop area. Quarantines are imposed in order to inspect incoming goods, detect attempted entry, and thus screen out potential pests. Quarantines operate under federal laws at various ports of entry into the United States, and also under state laws instituted to prevent interstate spread of unwanted potential pests into new areas. Quarantines are regulatory tactics requiring adequately trained personnel and detailed surveillance. Thousands of potentially damaging pests have been intercepted and excluded from the United States as a result of quarantines.

Eradication

The reduction of established pest populations in an area may be accomplished in a number of ways, all of which are eradicative in principle. Various cultural practices instituted for crop production admirably exemplify the principle. Cultivation exposes soil insects such as beetle larvae (grubs) to the lethal action of natural enemies and weather. Cultivation destroys subterranean habitats of pests. Sanitation practices are important eradicative measures; populations of pink bollworm, for example, can be substantially reduced if crop residue is destroyed by shredding, chopping, or burying of the old cotton stalks. Flea beetles, aphids, and other overwintering species are left homeless if weeds, border vegetation, and old crop residues are burned or buried. Early detection, roguing, and destruction of branches of trees infested with shot-hole borers or bark beetles is often sufficient to eradicate economically damaging populations (Metcalf et al., 1962). The threat of soil pests such as wireworms may be effectively reduced by sustained flooding of the soil before crops are planted. Celery and other vegetable crops are produced despite this pest in Florida muck soils after flooding (Guzman et al., 1973).

The threat of the granulate cutworm (*Feltia subterranea* Fab) to tomato production can be eliminated by staking the plants on a ground cover of plastic mulch (Price, 1974). The use of a mulch of aluminum foil to repel aphids successfully reduced the incidence of virus in vegetable and ornamental plants, but the "repulsiveness" of aluminum foil is questionable. Rather, the plants seem unattractive to aphids when grown on aluminum mulch or foil.

Soil pests are eradicated by treatment with steam, by fumigation, or by applications of insecticides to the soil before crops are planted. The use of insecticides sprayed on insect populations is perhaps the most common method of eradication in modern agriculture. Almost all kinds of arthropod pests, beetles, caterpillars, maggots, grubs, and mites, are suppressed by treatment with chemicals. Release of smoke-bomb fumigants in closed houses eradicates greenhouse insects.

Strip crops or harvests and crop-rotation techniques can suppress pests on companion crops. Lygus bugs were lured away from cotton by alternating cotton with strips of alfalfa, a more favorable host: the pest was thus eradicated from the cotton (Stern, 1969). Crop rotation replaces a host plant with another, nonhost plant in a field where pests have built up. For example, fields planted with corn, cotton, and soybeans in rotation effectively suppress the build-up of pests of any one of the three crops.

Therapy

The aim of therapy is to control a pest after it has begun its parasitic or predaceous relationship with its host, but before that potentially harmful relationship has resulted in damage to the host as an organism. Many crops are treated as seedlings with a variety of systemic granular pesticides (aldicarb, carbofuran, disulfoton, and phorate) applied to the soil. The toxicants are later incorporated into the plant through the roots and translocated to twigs and foliage. Sap-feeding insects that attack the plant absorb lethal quantities of the toxicant. Current recommendations for aphids, mites, leafminers, and other species include the use of systemic insecticides. Many pests of ornamental plants are controlled by curative applications of systemic chemicals.

Because of the selective pressure of such lethal chemotherapeutants on a population, there often develops, in time, a general increase in the level of tolerance of the chemical in the pest population. After multiple applications, the concentration required to kill may become higher. Consequently, routine treatment with therapeutic chemicals for many generations may result in populations that are difficult to control (Brown, 1968). Infrequent use of therapeutic materials to augment other means of population control is, therefore, more advantageous than long-term use of single curative agents.

Vertical Resistance

Entomologists often equate varietal resistance of crop plants with the term antibiosis, which is the direct deleterious effect of host plants on insects, causing death, reduced fecundity, or retarded development. Vertical resistance to insects, as to other plant pests, is the almost complete resistance of the crop-plant variety to the predominant strains of the pest. It is usually monogenic and expressed in an all-or-nothing effect. The resistance of many wheat varieties to Hessian fly appears to be vertical (Painter, 1951).

Horizontal Resistance

The broad polygenic resistance of many plants to insects or mites has been termed horizontal resistance. Most plants possess some degree of horizontal resistance to all arthropod pests; in fact, most plants are resistant to most pests. If this were not true, hordes of plant-feeding species would strip plants of any foliage, destroy the wood, and consume the root system. In crop plants, the rate at which populations increase or individuals develop varies from plant to plant.

The forms of horizontal resistance termed preference and tolerance are usually less well-defined than is vertical resistance (Painter, 1951). Preference is a characteristic of the pest. It refers to the discriminatory activities of a pest population, which, when provided with several choices, will prefer specific cultivars over others. Tolerance, a characteristic of the host plant, is the capacity to survive, grow, and produce in spite of the pest. Preference of aphids, thrips, and spider mites for different chrysanthemum cultivars grown under different horticultural practices and the tolerance of one cultivar (Iceberg) to spider-mite damage have been demonstrated (Poe and Green, 1974).

The corn earworm (Figure 8.3) feeds on many hosts, but its largest population and most rapid development occur on its primary host, corn. Secondary hosts include tomatoes, cotton, and sorghum. Among corn varieties, physical characteristics such as husk tightness, length of silk tunnels, and the presence of various physiological feeding components are revealed as limiting factors; these may slow population increase (Luckmann et al., 1964). Feeding on horizontally resistant cultivars reduces the breeding rate or amount and the number of offspring produced by a species. It may also prolong the life cycle by slowing the rate of development; this results in fewer generations per unit of time and gives an overall advantage for control (Leuck and Skinner, 1971).

Although the pest population itself may not be adversely affected by feeding on the host plant, and although no horizontal resistance is evident, there may be tolerance of insect or mite attack because of the vigor and general thriftiness of the host. Tolerance of some plants during one season of the year and not another appears to be due primarily to the physiological state of the host plant and its nutritional value to the pest population. Poe (1971) and Dabrowski et al. (1971) showed that spider-mite populations were more common and damaging on fruiting strawberry plants than on nonbearing plants. Pest responses to plant nutritional quality might reflect a change in the ratios, composition, or amounts of biochemical plant-growth regulators that govern production of leaves and fruit. Recognition of plant responses and of the capacity of the plant to tolerate damage (as well as of other regulating factors, such as preference of pests for certain cultivars over other cultivars) can help establish the relationship of population levels to economic thresholds. The pest-control specialist may not be able to rely on horizontal resistance as a tactic for the control of all species of insect and mite pests. Because of the selective nature of the pests themselves, most plants are attacked by only a few pest species.

The corn earworm, green-peach aphid, beet armyworm, granulate cutworm, and two-spotted spider mite have wide ranges of acceptable host plants. They can successfully complete development on many vegetable, agronomic, or weed hosts of several plant families. For example, the beet armyworm attacks corn and sorghum (Gramineae), tomatoes (Solanaceae), cotton (Malvaceae), chrysanthemum (Compositae), carnation (Carophyllaceae), and several other species (Poe et al., 1973). Pests are commonly restricted to a few related species or families of plants. *Keiferia lycopersicella* (Walsh), the tomato pinworm, seems restricted to plants in the family Solanaceae: tomato, potato, eggplant, and a few weeds of the genus *Solanum* (Batiste and Olson, 1973). Tomato is the most suitable host plant, with potato and eggplant poor seconds. But there is variation in host susceptibility among cultivars of tomatoes.

It is often possible, because of their feeding preferences, to group primary insect pests of the various plant families: pests of Cucurbitaceae, pests of Leguminosae, pests of Cruciferae. Horizontal resistance is evident, although unrecognized, among many common crop plants (Brett and Sullivan, 1974). Horizontal resistance of plants, which results in slowed development of pest populations, can be important in pest survival as well as in plant protection. Poor but sufficient hosts may retard growth, but would provide sources of food for pests when more desirable hosts are absent. From secondary hosts, pests may invade fields of crop plants when the population begins to develop rapidly. Because of the slowed rate of development on poor alternate hosts, the annual population cycle of the pest may be slowed.

Protection

Perhaps the most common use of chemical pesticides in agriculture is protective. Treatment of plants prophylactically by the application of chemical toxicants to prevent destruction is widely practiced, especially on high-unit-value commodities such as vegetables and ornamentals. Pesticide-use research by state agriculture experiment stations and recommendations by extension personnel often emphasize the principle of protecting the crop from pests. I remind the reader at this point that modern pest control involves the judicial—not routine or regular—application of pesticides. The pest-control specialist uses pesticides, but only when needed.

Wingless female spring canker worms, which overwinter in the soil as pupae, are prevented from reaching their host by the application of a band of sticky tanglefoot about the trunk of the tree (Metcalf et al., 1962). Lara (1970) protected bananas on the tree by encasing the bunch in insecticide-impregnated plastic bags. Protective measures are common in glasshouses or in closed-structure farming, where areas can be screened tightly to prevent entry of many pest species. Moths, beetles, and even aphids and thrips can be denied entry into houses by adequate screening of vents and other openings. Plant material, supplies, or equipment introduced into the house must be free of vagabond pests, as must the workers who tend the plants. Protection of corn from migrating chinch bugs was achieved by a creosote-treated furrow or paper barriers erected around the border of the fields. Similarly, Coulee or Mormon cricket invasions may be halted by a short metal fence (Metcalf et al., 1962).

Storage of grains and food surplus is done with protection from pests in mind. Preservation as frozen, canned, pickled, salted, or gassed products are all means of protecting our crops from pest destruction.

Avoidance

Knowledge of insect life histories and annual population buildup has provided a tactic for avoidance of large population numbers and damage to crop plants. Where plants can be grown in an environment largely free of pest species, undue damage can often be avoided. Numerous pest species overwinter in low numbers and begin to increase with warm spring weather and the onset of the annual crop cycle. Establishment of crop plants in the field before pest populations intensify, or the use of early and rapidly maturing varieties, can have a profound effect on the amount of pest damage sustained. Most common are the planting schemes developed for wheat, which enable growers to avoid destructive populations of the Hessian fly. The recommendations include relatively late fall planting (after adult flies have oviposited and died), sanitation, and burial of wheat stubble Metcalf et al., 1962). The range of possibilities for avoidance is limited, however, to certain areas and to those insects that have well-defined life cycles during the year; it is also restricted to those crops that have only one or two key pest species. Menke (1973) showed by simulation that most soybean damage by the velvet-bean caterpillar could be avoided by planting earlier; this enabled the plants to produce before the normal generation increases of caterpillars reached high levels. Also, early planting permits early-maturing varieties of cotton to avoid damage from the bollweevil or pink bollworm.

INTEGRATED PEST CONTROL

The modern science and art of invertebrate pest control has shifted toward combining the best of several techniques for plant-pest control. Similarly, the crop-protective efforts by vermin-control specialists, weed scientists, nematologists, plant pathologists, and entomologists can no longer be applied independently. Instead, there has been and will continue to be an integration of efforts for pest control by these specialists with those of specialists in basic production sciences (horticulture and agronomy) and in economics. Concern over adequate food production, energy, and other natural resources, pesticide misuse, and pollution has led to an integration of the principles of pest control from each segment of crop production. We may expect, now and in the future, not only integrated techniques for the control of specific pests but also a totally integrated program, with contributions from all segments of agriculture. This will enable scientists to make appropriate and advantageous trade-offs among production alternatives and ensure an economically produced crop, one free of intolerable pest damage in a healthy environment.

LITERATURE CITED

Batiste, W. C., and W. H. Olson. 1973. "Laboratory evaluation of some solanaceous plants as possible hosts for tomato pinworm." *J. Econ. Entomol.*, 66:109–111.

Borror, D. J., and D. M. Delong. 1971. *An Introduction to the Study of Insects*, 3rd ed. New York: Holt, Rinehart and Winston.

Brett, C. H., and M. J. Sullivan. 1974. "The use of resistant varieties and other cultural practices for control of insects on crucifers in North Carolina." *North Carolina Agr. Expt. Sta. Bull.*, no. 449.

Brown, A. W. A. 1968. "Insecticide resistance comes of age." *Bull. Entomol. Soc. Amer.*, 14:3–9.

Chapman, R. F. 1971. *The Insects: Structure and Function.* New York: Elsevier.

Clark, L. R., P. W. Geior, R. D. Hughes, and R. F. Morris. 1967. *The Ecology of Insect Populations in Theory and Practice.* New York: Wiley.

Dabrowski, Z. T., J. G. Rodriguez, and C. E. Chaplin. 1971. "Studies on the resistance of strawberries to mites, IV: Effect of season on preference and nonpreference of strawberries to *Tetranychus urticae*." *J. Econ. Entomol.*, 64:806–809.

Guzman, V. L., H. W. Burdine, E. D. Harris, Jr., J. R. Orsenigo, R. K. Showalter, P. L. Thayer, J. A. Winchester, E. A. Wolf, R. D. Berger, W. G. Genung, and T. Z. Zitter. 1973. "Celery production on organic soils of south Florida." *Fla. Agr. Expt. Sta. Bull.*, no. 757.

Knipling. E. F. 1955. "Possibilities of insect control or eradication through the use of sexually sterile males." *J. Econ. Entomol.*, 48:459–462.

Lara, F. 1970. *Problemas y Procedimientos Bananeros en la Zona Atlantica de Costa Rica*. San Jose, Costa Rica: Imprenta Trejos Hnos.

Leuck, D. B., and J. L. Skinner. 1971. "Resistance in peanut foliage influencing fall armyworm control." *J. Econ. Entomol.*, 64:148–150.

Luckmann, W. H., A. M. Rhodes, and E. V. Wann. 1964. "Silk balling and other factors associated with resistance of corn to corn earworm." *J. Econ. Entomol.*, 57:778–779.

Martin, H. 1973. *The Scientific Principles of Crop Protection*, 6th ed. New York: Crane Russack.

Menke, W. W. 1973. "A computer simulation model: The velvetbean caterpillar in the soybean agroecosystem." *Fla. Entomol.*, 56:92–102.

Metcalf, C. L., W. P. Flint, and R. L. Metcalf. 1962. *Destructive and Useful Insects*. New York: McGraw-Hill.

Metcalf, R. L. and W. H. Luckmann, eds. 1975. *Introduction to Insect-Pest Management*. New York: Wiley.

Painter, R. H. 1951. *Insect Resistance in Crop Plants*. New York: Macmillan.

Pfadt, R. E. 1971. *Fundamentals of Applied Ecology*, 2nd ed. New York: Macmillan.

Poe, S. L. 1971. "Influence of host-plant physiology on populations of *Tetranychus urticae* Koch (Acarina: Tetranychidae) infesting strawberry plants in peninsular Florida." *Fla. Entomol.*, 54:183–186.

———, C. L. Crane, and D. Cooper. 1973. "Bionomics of *Spodoptera exigua* (Hub.), the beet armyworm, in relation to floral crops." *Proc. Amer. Soc. Hort. Sci. Trop. Reg.*, 17:389–396.

———, and J. L. Green. 1974. "Pest-management determinant factors in chrysanthemum culture." *Proc. Fla. Hort. Soc.*, 87:467–471.

Price, J. F. 1974. Effects of tomato cultural systems on insect populations and on fruit production. M.S. Thesis, Univ. Florida, Gainesville.

Stern, V. M. 1969, "Interplanting alfalfa in cotton to control lygus bugs and other insect pests." *Proc. Tall Timbers Conf. Ecological Animal Control by Habitat Management* 1:55–70.

Wilson, E. O., and W. H. Bossert. 1971. *A Primer of Population Biology*. Stamford, CT: Sinauer Associates.

9

Weeds and Their Control

Earl G. Rodgers

Weeds are plants growing where they are not wanted. Since the beginning of recorded history, weeds have demonstrated aggressive characteristics and growth that have interfered with human activities in many different ways. Accordingly, human beings have exerted extensive efforts to control the growth and spread of these plants. According to Martin et al. (1976), the destruction of weeds was established as one of the original features of crop production before the dawn of recorded history. Throughout the ages, people have been forced to control weeds if their food crops were to produce acceptable yields. Weeds interfere also with many other human activities. Various species of weeds produce prolific growth in aquatic sites, thereby interfering with transportation on waterways as well as with aquatic animal life and recreational activities. Weeds interfere with all phases of domestic-plant production, whether it be for food, feed, or fiber.

Weeds, unlike all other kinds of pests, are similar to the plants they injure. Domesticated plants are almost always higher-order plants. Weeds likewise are higher-order plants and, therefore, have the same general growth requirements as the plants with which they compete (Chapters 2 and 3). In the production of a given crop, a common practice is to ensure an adequate supply of nutrients and water to the crop plants. Supply of these basic requirements also benefits weeds that grow adjacent to the crop plants and compete with them for these basic requirements. Weed-control procedures, then, must have the

narrow margin of safety required to protect the crop plants while providing control of the similar weed plant.

The plant community is a complex association. Under a particular
set of environmental conditions, the natural progression is toward a
climax vegetation. Muzik (1970) has shown that human attempts to
change this natural progression in order to grow crops is, in effect, an
uphill battle against the natural succession of plant growth. For example, certain climates favor trees or brush instead of grasses. In these
areas, growing grasses along the roadside is very difficult, since the
natural ecology leads to woody plants as the climax vegetation. In
other areas, grasses are favored and trees are difficult to establish. We
have been forced to adapt our agriculture to the climatic conditions of
a given region. Thus, we have crop belts in which certain crops, such
as specific grains or fruits, are best adapted and are produced best. No
major crop species would survive long without human help, however,
since they are not as well-adapted to the environment as are certain
other plants that compete with crop plants and that would eventually
take over the fields (Figure 9.1). Such competitive plants are weeds.

FIGURE   9.1

A field of peanuts 16 weeks after application of a selective herbicide. Florida pusley and
yellow nutsedge are dominant in the untreated bed (center); Florida beggarweed
dominates the treated beds (left and right). [Courtesy of E. W. Hauser.]

The number of weed species that populate the Earth is not known, and in fact, cannot be calculated, because a given species may be desirable and, accordingly, not a weed under some circumstances but undesirable and, therefore, a weed under different circumstances. An example is the blackberry (*Rubus* spp.), which is a cultivated crop plant and not a weed when grown for fruit production. But the same plant grows prolifically on many cultivated lands and requires continued efforts for its control. Under such conditions, this plant then becomes a weed. Furthermore, volunteer hills of peanuts (*Arachis hypogaea* L.) or plants of some other crop species that may appear in a field of solid corn (*Zea mays* L.) are considered weeds under those conditions.

CROP LOSSES TO WEEDS

Most plants that appear and persist without human aid (and often despite our efforts) are considered weeds. The losses attributed to weeds are widespread and of great magnitude. Irving and LeClerg (1965) and Crafts (1975), among others, estimated that the annual cost of weeds to agriculture in the United States amounted to at least $5 billion in 1965. Buchanan (1975) estimated that this loss in dollars that are corrected for inflation likely doubled, to $10 billion, by 1975. Approximately half this total loss arises from the reduction in yield of agricultural products that is caused by the competition of weeds. The other half of this loss reflects the cost of efforts to prevent this loss of crop yields from becoming even greater.

The kinds of agricultural losses due to weeds include: reduction in quantity and quality of crops, pastures, meat, milk, and wool; increased cost of crop production because of cultivation, hoeing, mowing, and buying and spraying herbicides; limitation of crop choice, because some crops will not grow well in areas that are heavily infested with certain weed species; reduced land values; increased harvesting costs; lethal effects on livestock by poisonous weeds; and the harboring of insects and pathogens that attack and injure crop plants.

In addition to these losses to agriculture, some weeds cause such human discomforts as skin determatitis (poison ivy, *Rhus radicans* L; and poison oak, *Rhus toxicodendron* L.) and hay fever (pollen from ragweed, *Ambrosia* spp. and many other species). Also, railroads, public utilities, and highway departments spend vast sums of money for weed control on their rights of way, and staggering sums of money are spent annually to control weeds in aquatic sites.

BENEFITS FROM PLANTS USUALLY TERMED WEEDS

Notwithstanding the tremendous losses attributable to weeds, some plant species commonly considered weeds have some individual advantages. These advantages usually are not important enough to warrant the use of cultural practices to preserve these plants, but such plants in natural stands should be used however possible for human benefit. Andersen (1968), for example, pointed out that weeds may be grown for experimental use in order to develop effective control procedures. Some benefits derived from plant species that are often considered weeds can be enumerated: they provide organic matter to the soil; they prevent leaching of nutrients and return nutrients to the soil when they die and decompose; they provide esthetic value as covering for waste places; they control erosion; many provide pasture and hay for livestock; a few provide edible fruit; some serve medicinal purposes (bindweed, *Convolvulus* spp., roots produce an extract useful in the control of blood clotting); and, finally, some provide germ plasm useful in plant-breeding programs.

CLASSIFICATION OF EFFORTS
TO REDUCE WEED POPULATIONS

The numerous terms used for efforts to prevent, retard, or eliminate populations of agricultural pests other than weeds often have little or no utility in weed science. Notwithstanding the severity and widespread distribution of weeds, efforts to negate their adverse effects have been classified into the three categories of prevention, control, and eradication. These three terms have become an integral part of the literature of weed science, and no changes in their usage are recommended here. But these terms, it seems, should be related to the terminology of plant-pest control used elsewhere in this book. "Prevention" is the equivalent of "exclusion," and "control" and "eradication" as used in this chapter are examples of "eradication" as defined in Chapter 5.

Weed Prevention (Exclusion)

Weed prevention is the term used by weed scientists for denying a weed species access to a given area. Because weeds are transported as seed and vegetative parts across boundaries of various kinds, cooperative efforts in prevention are needed in order to achieve significant results. Accordingly, some countries and other geographical entities have established exclusionary programs designed to be effective within their boundaries. State, national, and international efforts often are essential, however, if maximum results are to be realized in weed prevention.

Despite the numerous weed species already present in the United States, recent plant surveys have detected at least 1,200 additional species of potentially serious weeds in other countries (Reed, 1977). Prior to 1975, the only deterrents to the introduction of these plant species into the United States were portions of the Federal Seed Act designed to prohibit introduction of weed seed or other vegetative reproductive plant parts with imported crop seed. Any plant or plant part could have been introduced into the United States without legal restrictions prior to January 1975, as long as the plant being brought in was not an actual or suspected host to a disease-causing organism, insect, or other pest. However, after long efforts by many people, a Federal Noxious Weed Act was signed by President Ford on January 4, 1976. This act authorized the Secretary of Agriculture to develop a weed-inspection program at points of importation to prevent plants classified as serious weeds from being introduced into this country. The enactment of this legislation has contributed strongly to the preventive (exclusionary) weed control program for the United States.

Weed Control (Eradicative Measures)

Weed control, in the sense intended by the weed scientist, means destroying enough weeds that the remaining weed plants cannot compete significantly with domesticated plants or interfere with other human activities. Control becomes more difficult and more expensive as more weed species are allowed entrance into an area. Conversely, less control and, accordingly, lower costs are required where effective weed-prevention programs have been carried out.

Weed Eradication (in the Literal Sense)

Eradication, in the sense intended by the weed scientist, means the complete elimination of a weed species (including those in and on the soil) and all vegetative parts. Eradication in this sense is seldom justified, because the costs are much higher than for control as defined in the preceding section. Procedures for both control and eradication are similar, in that both destroy existing weeds. (Control and eradication as used in this chapter thus exemplify eradicative measures as defined in Chapter 5.) However, eradication, in the sense intended by the weed scientist, destroys *all* the weeds, whereas control destroys *some* of the weeds, enough to make damaging competition or interference by the remaining weeds impossible. In practice, destroying all the weeds costs much more than destroying just enough weeds to control their population. These added costs are seldom justified, except in intensive crop-production programs or in small areas managed for highly specialized purposes, where costs are relatively unimportant. Occasionally, the extra costs may be justified in order to eradicate small infestations of a highly noxious perennial weed that could spread to a much larger area if left undisturbed. Eradication of incipient infestations of new weeds near entrance avenues would contribute effectively to prevention programs designed to deny the entry of new weeds into an area.

WEED-CONTROL PROCEDURES

Weeds should be controlled by the least-expensive available technology that does not interfere with other phases of crop production or other human activities. Any control practice should be used only when its results are expected to be more economically advantageous than the results of not using any control. The primary weed-control procedures can be classified into four major categories of methods: physical, managerial, biological, and chemical. Under some circumstances, a single procedure may be satisfactory, but most conditions require use of two or more different methods in order to produce a crop or clear a weed from a particular site at the lowest possible cost.

Physical Procedures

Hand Pulling and Hoeing

Removal of weeds by hand was probably the first means of combating undesirable plants. This method is used today to remove weeds growing close to valuable plants. Hand pulling is practiced most commonly in plantings of ornamental crops, and in other highly intensive plant-production programs. Current high prices and scarcity of labor prevent extensive use of this method. Use of the hoe is also limited today by the high price and scarcity of labor in the United States. Hoeing requires much human effort, but may sometimes be necessary to remove weeds in crop rows and other areas where other weed-control procedures are not satisfactory.

Tillage and Cultivation

Stirring the soil for the benefit of domesticated plants is known as tillage. Tillage performed after the crops are planted is commonly known as cultivation. Annual weeds are controlled more effectively by tillage than are perennials. Tillage for seedbed preparation is intended to destroy existing weeds and to stimulate germination of weed seed in the topsoil layers, so that resulting weed seedlings can be killed by disking or harrowing before the crop is planted. Cultivation is intended to kill weeds that are growing between crop rows after emergence of the crop plants and also to reduce the population of weed-seedlings in the row by covering them with soil. Cultivation after a crop is planted but before the crop seedlings have emerged is often practiced uniformly over a field to kill young weed seedlings and to allow the crop seedlings to begin growth before the weeds, thereby giving the crop plant a competitive advantage. This practice is called blind cultivation, and is commonly done with a spiked tooth harrow, rotary hoe, or other implements with finger-like tines that lightly penetrate the soil.

Shallow cultivation is usually best. Deep cultivation is no better than shallow and sometimes not as good, particularly with weeds like crabgrass (*Digitaria* spp.), with which it is desirable to remove most soil from the plant roots. Furthermore, deep cultivation turns up weed seed that were dormant at the greater depth and thus causes additional weeds to appear.

Mowing and Chopping

These practices of weed control are most common in sodded areas, such as pastures, lawns, and waste places. Complete destruction of vegetation above a certain height is provided. Mowing simply severs the weeds at this predetermined height, and all of the plant above this height may or may not be pulverized into small particles, depending upon the kind of mower used. Weeds not pulverized into small pieces deteriorate slowly, and may provide excessive shade over the pasture species. Therefore, mowers that cut the severed plants into small parts are preferred in pasture and turf areas. Choppping cuts the weeds at about the same height as mowers, and cuts the severed portion of the plant into sections four to six inches long. This method is good for use in pastures and waste areas, as well as for destroying stands of woody plants with stems of relatively small diameter. Chopping is not usually acceptable for use in turf, because the chopper blades often penetrate the soil and leave open slits that are unsightly and contribute to excessive loss of soil moisture.

Mowing and chopping allow small, uncut weeds to continue growth and possibly produce a seed crop. Furthermore, root systems are not destroyed, and they may support developing tillers from the old stubble.

Burning

Burning is of limited value as a method of weed control. Burning may be used to kill weeds in a vegetative state, but weed seed generally are not injured seriously by this procedure unless sufficient debris is present to provide intense heat. Burning may cover an entire area, or small patches of weeds may be burned with a flame gun. Only controlled burning should be practiced, and then only if not prohibited by law.

Flaming

Flaming exposes tender weed seedlings very briefly to an intense flame to produce a searing effect. It has been used with success in cotton (*Gossypium hirsutum* L.) and a few other crops that have a thick epidermis near the base of the stem. A series of small burners, each similar to but smaller than an acetylene torch, is attached to the foot of a cultivator that moves parallel to, but a few inches away from, the row. The burners direct the individual small flames into the crop row from both sides at ground level as the cultivator is moved along. The flames kill the young tender weeds in a row without injury to the older, more resistant crop plants.

Flooding

In regions where irrigation is practiced, flooding is sometimes used to control weeds. Keeping an area covered with water reduces the aeration that some dormant weed seed and the roots of some living plants need to live. Flooding may be continued under suitable circumstances until these seed and plants are dead. Limited research data, however, suggest that some weed seed can remain viable after submersion for several months or even years. This method is satisfactory in cropped areas only with such crop plants as rice (*Oryza sativa* L.), which can tolerate excessive moisture.

Managerial Procedures

Managerial practices of the grower often influence the selection of weed-control procedures and the success of a given procedure after it has been selected. Methods should be selected to give the crop plant a specific advantage over the weed.

Crop Rotation

In almost all agricultural areas, the continuous production of a particular crop on a given area of land will increase the population of those weed species that compete well with the particular crop being grown. If different crops with varying growth habits are grown on a given area of land, no special kinds of weeds would have an opportunity to become better established than any other kinds. Weeds and crop plants that are morphologically similar tend to grow together more compatibly. Therefore, a rotation of crops with different growth habits avoids favoring any one group of weeds that may grow well with any one crop in the rotation.

Taking Advantage of Differences between Crop and Weed

Specific behavioral differences between the crop and weed plants should be used to create an advantage for the crop plants whenever possible. There are several such differences that are bases for effective weed control. For example, crop plants that mature early are ideal for growing where weeds mature later; the weeds then can be destroyed by physical means before they produce seed.

Time of Seed Germination

Crop and weed seedlings normally enter a stage of rapid growth and development soon after their emergence. Weed seed of a particular species can be expected to germinate and develop seedlings at a particular season of the year. For example, if the major weed species on a piece of land are known to germinate in late spring and thereby become most problematic at that time, crops should be planted on that area early in the spring, so that the crop plants would be well into their growing period and therefore have a distinct competitive advantage over the weed seedlings that appear later.

Rate of Growth

Crop varieties that are well-adapted to an environment grow faster and compete better against weeds than varieties that are less well-adapted. Such crop plants, especially those that grow erect, quickly shade the ground and deny adequate light to weed seedlings.

Season of Growth

Those crops that grow during a season when weeds are not growing would be avoiding competition with weeds that develop during other seasons on the same land. For example, crops grown in the winter, as in tropical and subtropical regions, are not in competition with weeds that grow in the summer.

Soil Management

The soil may be managed in ways that aid in weed control. Seedbed preparation should be as shallow as is compatible with normal crop-planting operations and seedling development. Stirring the soil to a greater depth brings up dormant weed seed from these depths to more shallow soil zones, thereby allowing weed seed to germinate and compete with the crop seedlings.

Minimum tillage may be carried out in conjunction with the production of selected crops. This practice involves planting the crop seed with the minimum disturbance of the existing vegetative soil cover, which had been retained to protect the soil surface from erosion. This existing vegetative cover, however, is killed, most commonly by means of a contact herbicide sprayed uniformly over the area at the time the crop seed are planted. The dead residue of the vegetative cover then protects the soil from erosion, and since the soil is not stirred, dormant weed seed are not brought to the surface.

Cropping Practices

The cropping practices used in agriculture affect the growth and development of weeds in a given area. Row crops often leave much of the soil surface unoccupied by crop plants, and weeds may appear and abound between the rows and between the plants in the rows. In orchards, weeds frequently serve to control soil erosion between the rows of trees (Figure 9.2).

Broadcast or close-drilled crops, however, cover the soil more fully; sod is even more thorough in covering all the soil surface and is more efficient, therefore, in preventing weed growth. Nevertheless, continued care must be exercised to keep turf free of weeds (Figures 9.3 and 9.4).

Crop-plant populations should be as large as can be supported by environmental conditions. The more crop plants there are per unit area of soil surface, the more competitive advantage they will have over potential or existing weeds. Increased crop-plant densities can be arranged by adjusting distances between adjacent rows as well as by the spacing of plants within the row.

FIGURE 9.2
A peach orchard in which weeds have been controlled under tree canopies but allowed to grow between the rows to prevent soil erosion. [Courtesy of the Weed Science Society of America.]

Harvesting and grazing methods used on pastures also directly affect the numbers of weeds. Overgrazing probably is the most frequent cause of weedy pastures. Overgrazing depletes the food reserves in the roots of pasture plants and reduces the density of the pasture sod, thereby reducing the ability of these plants to compete satisfactorily with invading weeds.

Biological Procedures

The reduction of the population of a plant or animal species by means of natural enemies such as parasites, predators, and pathogens, is known as biological control. This is a natural process that has been continuous since the origin of land plants. Planned use of this method to control weeds, however, is fairly recent. Intensive efforts to direct the injurious effects of these natural enemies toward the control of weeds have met with varying degrees of success.

FIGURE   9.3

A lawn with a weed-free area (center) that had been treated with an herbicide and surrounding weed-infested areas that had not been treated. [Courtesy of the Weed Science Society of America.]

The greatest success in biological weed-control efforts has been in the use of insects to control selected weed species. Classic examples of such success include: the control of lantana (*Lantana camara* L.) by caterpillars of *Plusia verticillata*, and by larvae of the tortricid moth (*Crocidosema lantana*), of the seed fly (*Agromyza lantanae*), and of lycaenid butterflies (*Thecla echion* and *T. bazochi*) in the Hawaiian Islands at about the turn of this century (Crafts, 1975); the control of prickly pear (*Opuntia* spp.), primarily by the cactus moth (*Cactoblastis cactorum*), in New Zealand and Australia in the 1920's (Klingman and Ashton, 1975); and the control of St. John's wort (*Hypericum perforatum* L.) by goatweed beetles (*Chrysolina hyperici* and *C. gemellata*) on ranges of the western United States in the 1950's (Muzik, 1970). Intensive efforts have been recently made to use additional species of insects, as well as various fungi and viruses, as agents for biological control of weeds.

The greatest success seems to come from selecting organisms that are parasitic to the weed in the geographic region where the weed

FIGURE   9.4
Application of an herbicide to control weeds in a cemetery. [Courtesy of the Weed Science Society of America.]

itself originated, and introducing them into the current habitat of the weed. In introducing a parasite to control the weed in this new area, however, extreme care and extensive testing are essential to ensure that the parasite is sufficiently specific to the weed to be controlled; otherwise the parasite might eventually spread to related crop plants and become a pest to them. To be considered safe, the introduced parasite must demonstrate a "preference for death" over a change to a food source other than the weed species that is to be controlled.

Weeds that can most likely be controlled by biological procedures are individual species that predominate a relatively large area that has not been much disturbed by human activities. The weed should also be a perennial, so that the parasitic organisms will have a source of food throughout the year. Annual weed plants on crop lands, therefore, are unlikely to be controlled by biological control procedures.

Herbicidal Procedures

Chemicals used for weed control are termed herbicides. The use of chemicals to control weeds is not a new practice; some compounds, such as sodium chloride and sulfuric acid, were used during the late nineteenth century to kill unwanted vegetation. The intensive use of herbicides for weed control, however, began around 1945 with the discovery of the selective phytotoxicity of 2,4-D [(2,4-dichloro-phenoxy)acetic acid]. This compound kills plants of some species with little or no injury to others. This characteristic stimulated much interest among weed scientists, and research soon thereafter demonstrated that this selectivity allowed use of 2,4-D to control many dicotyledonous weeds without injury to grass crops such as corn and other grains. Subsequent research demonstrated that treatment with 2,4-D or some other herbicide might cause highly variable plant responses. This variability was soon traced to such sources as different stages of plant growth, different environmental conditions at the time of application, and, particularly, different rates at which the herbicide was applied.

No single herbicide treatment can provide satisfactory control of weeds under all circumstances. This fact demonstrated a need for additional herbicides capable of controlling weeds not killed by compounds used earlier. Accordingly, biochemists soon developed new herbicides. Because of a simultaneous decrease in the labor supply and increase in the cost of available labor for physical methods of weed control, the demand for herbicides increased, especially in the United States. The response to this demand has been dramatic. Only about 15 primary herbicides were available in 1950. Two decades later, more than 180 primary herbicides and at least 6,000 formulated products were available. The acreage treated with herbicides between 1959 and 1965 in the United States rose from 53 to 120 million, an increase of 126 percent (Ashton and Crafts, 1973). Shaw (1974) stated that herbicides were used in 1973 on more than 150 million acres, or about 50 percent of the total harvested crop acreage in the United States.

These rapid advances reflect the efficiency and economy of controlling weeds with herbicides despite the tremendous cost of herbicide development. Green (1973) pointed out the average research cost of each new pesticide was $5.5 million from synthesis through production of early permit materials. Cash flow is all outgo during the first six years, and an average period of 10.3 years is required before income from sale of the product equals the amount spent on its development.

Herbicides are classified into specific categories for convenience in referring to them. Each classification group has one or more identifying characteristics that determine which individual herbicides should be included. Since the basis for each classification is different, a given herbicide has a place in each system. Some of the primary classification groups are considered next.

Chemical Characteristics of Herbicides

Organic compounds contain carbon. Most important herbicides are organic, and may be grouped by their chemical families. Some major families are the carbamates, including metham (sodium methyldithiocarbamate) and propham (isopropyl carbanilate); the nitriles, including bromoxynil (3,5-dibromo-4-hydroxybenzonitrile) and dichlobenil (2,6-dichlorobenzonitrile); the phenoxys, including 2,4-D and 2,4,5-T [(2,4,5-trichlorophenoxy)acetic acid]; the triazines, including atrazine [2-chloro-4-(ethyl-amino)-6-(isopropylamino)-s-triazine] and simazine [2-chloro-4,6-bis(ethylamino)-s-triazine]; and the ureas, including diuron [3-(3,4-dichlorophenyl)-1,1-dimethylurea] and monuron [3-(p-chlorophenyl)-1,1-dimethylurea]. This list is not complete. Each of these families also contains economically important compounds that were not given as examples. There are several other major families, in each of which are several herbicides that are in extensive use. In addition, some important organic herbicides are not classified according to family; they are individual representatives of a specific pattern of chemical structure. An example of an important unclassified compound is bensulide [0,0-diisopropyl phosphorodithiote S-ester with N-(2-mercaptoethyl)benzenesulfonamide].

Inorganic compounds do not contain carbon. They have served important functions in the development of weed science, and some continue to be significant for specific uses. Most of the inorganic herbicides now in common use are salts, but a few acids, particularly arsenic compounds, have specialized uses.

Formulations of Herbicides

Herbicides are usually available in the form of a liquid or a wettable powder. Either form of the compound is applied as a spray, with water used most commonly as the diluent. Some herbicides are also formulated as granules to prevent the hazard of drift that exists when materials are applied as a spray. Granules provide longer activation periods than do spray applications.

Response of Plants to Herbicides

Selective Herbicides.   Selective herbicides are those to which different plant species show a wide range of responses. One species may be killed, whereas another may not be injured by a particular application of the herbicide (Figure 9.5). Selectivity allows use of a particular herbicide to remove weeds growing among crop plants; the herbicide is applied to all plants uniformly, but the crop plants are not injured, whereas the weeds are killed. Many herbicides show selectivity, but 2,4-D is among the most widely known for its relative toxicity to broadleaf weeds and harmlessness to narrowleaf plants. That characteristic enables 2,4-D to serve many uses, among the more notable of which is killing ragweed growing in corn fields. Other herbicides show selectivity for other groups of plants.

The spectrum of weed species susceptible to a given herbicide is an important aspect of selectivity (Klingman, 1975). A narrow-spectrum herbicide is toxic to only one or a very few species of plants, and all other plants are tolerant. A broad-spectrum herbicide kills all plant species except a few. A narrow-spectrum herbicide has relatively little value in a program designed to control many weed species. These herbicides are justified only when a particularly noxious weed is to be

FIGURE   9.5

Peanuts (center) growing where weeds had been controlled by soil injection of an herbicide at the time of planting, by a preemergence treatment with an herbicide, and by one cultivation. Texas panicum, prickly side, sicklepod, and morning glory dominate the untreated plots (left and right). [Courtesy of E. W. Hauser.]

controlled; examples would include purple nutsedge (*Cyperus rotundus* L.) or Johnson grass (*Sorghum halepense* (L.) Pers.) on cultivated lands, or water hyacinth (*Eichornia crassipes* (Marti.) Solms) in aquatic sites.

A broad-spectrum herbicide is used for simultaneous control of many weed species, as commonly exist in a typical plant community. Since each major crop normally grows in competition with numerous weed species, the herbicide should have a spectrum sufficiently broad to cover a majority of the weeds. Many broad-spectrum herbicides are used in combinations of two or more to extend even further the spectrum of weed control. Almost all herbicides in common use are broad-spectrum.

Selectivity is perhaps a misnomer, because the herbicide does not "select" the plants it should kill among those being treated. Instead, phytotoxicity is based on the different responses of individual plants to that herbicide. Some plants tolerate a given herbicidal treatment very well and others do not. Such tolerance is based largely on morphological, physiological, and genetic plant characteristics.

Most selective herbicides are translocated within the plant from the point of entry to the point of accumulation and ultimate toxicity. Such materials do not commonly cause immediate phytotoxicity. They are absorbed and move some distance inside the plant, away from the point of entry, before becoming toxic. These materials often accumulate in the root, stem, or leaf-growing point, or in some other specific location inside the plant. Such points of accumulation most commonly are areas of high meristematic activity. When the accumulation reaches a certain level, toxicity occurs and the treated plant ultimately dies.

Contact Herbicides.   Contact herbicides are toxic to the plant protoplasm, and kill only those parts of the plant they contact. These compounds are not translocated, because the protoplasm is killed before translocation begins. Contact herbicides are usually nonselective, but a few show some selectivity, depending on the plant part that is contacted. Paraquat (1,1'-dimethyl-4,4'-bipyridinium ion), for example, is highly toxic to plant parts that contain chlorophyll, but has few or no ill effects on plant stems or on parts of the plant that do not contain chlorophyll.

Herbicide Application

Herbicides are applied as sprays to the foliage of growing plants or to the soil surface before or shortly after planting. Herbicides that are absorbed into the plant system through the leaves are applied when the plants are in an active stage of growth (Figure 9.6). Weeds are most easily killed by foliar sprays during the early postemergence growth period, but applications may be made at any time after plant emergence, depending on specific objectives. For example, herbicidal sprays may be applied during early stages of seedling growth to eliminate weed competition with crop plants; or an herbicide may be applied in corn fields as this crop is approaching maturity to eliminate various weeds that would interfere with mechanical harvesting procedures. Foliar sprays are termed postemergence applications.

Herbicides that affect the plant primarily through the root or that kill dormant or germinating seed are applied to the soil. Many of these herbicides have little or no effect on plants if applied only to the foliage. Some herbicides that exhibit primarily contact toxicity when applied to the foliage develop more selectivity when applied to the soil. For example, dinoseb (2-*sec*-butyl-4,6-dinitrophenol) exhibits contact toxicity to plant foliage, but it exhibits selectivity when applied before emergence; it thus allows satisfactory emergence of some important crop plants. such as peanuts, but kills the germinating

FIGURE   9.6

Application of an herbicide to corn after its emergence. [Courtesy of E. L. Knake.]

seedlings of many annual weeds. These materials can also be applied to the soil at different times relative to plant emergence.

Herbicides applied to the soil surface before the crop is planted are known as preplant treatments. Herbicides applied in this manner are most commonly toxic to weed seed that are in a dormant or germinating condition in the soil. The materials are usually toxic to crop seed or seedlings also, but applications before the crop is planted avoid injury to crop seedlings. Most herbicides applied before planting are incorporated, by disking or similar means, into the soil to a depth of about six inches (Figure 9.7). Such incorporation increases the probability that the herbicide will come into contact with the weed seed to be killed, and also reduces the losses due to volatility, which would be great for many compounds, such as EPTC (S-ethyl dipropylthiocarbamate) if they were left on the soil surface.

Preemergence herbicides are those applied to the soil surface after the crop seed are planted but before crop and weed seedlings have emerged. Being applied to soil devoid of all vegetatively growing

FIGURE   9.7

A volatile herbicide being incorporated into the soil before a crop is planted. [Courtesy of E. L. Knake.]

plants, these materials remain in the topmost layer of soil. As the emerging weed seedling approaches the soil surface, it comes into contact with the herbicide, absorbs a toxic quantity, and dies without emerging (Figure 9.8). Selectivity is essential, so that the crop seedling does not suffer such a fate; the crop seedling must be able to penetrate the treated layer of surface soil without injury. If the herbicide is leached by rainfall into contact with the weed and crop seed below the soil surface, this same selectivity must prevail. The weed seed should be susceptible to injury by the herbicide, and the crop seed must be tolerant. Without this selectivity, herbicides cannot be used successfully for preemergence treatment.

Various soil surfaces should be maintained free of all vegetation for indefinite periods of time. Herbicides that are toxic to both dormant seed and roots of living plants, and that have relatively long residual effects in the soil, provide such maintenance most effectively. Several

FIGURE   9.8

Peanuts in a field where a preemergence herbicide had been applied along the crop rows (left and right); soil in the center had not been treated. Cultivation will control the weeds growing between the herbicide-treated rows. [Courtesy of J. F. Miller.]

herbicides, including monuron and diuron, have these characteristics. Such herbicides should be held strongly by soil particles in the topsoil horizons, to reduce excessive loss by leaching and to avoid lateral movement to adjacent areas where the herbicide toxicity may cause undesirable effects. Examples of such areas include permanent fence rows (Figure 9.9), lumber yards, and various industrial sites. These herbicides may be applied directly to the soil surface or to vegetation to be killed.

INTEGRATED WEED CONTROL

Because almost all weed-control efforts are directed toward many weed species, which often exhibit highly diverse growth habits and characteristics, more than one control method is commonly required to maintain weed populations below the economic threshold. The best method for each set of circumstances should be selected. However, the best control of weeds most often results from the use of two or more different methods in appropriate sequence or combination. The use of multiple methods of weed control in a given area or crop is more common than use of a single method alone. For example, tillage of the soil is practiced primarily for weed control, but the use of an herbicide as an additional method of weed control in the same area, either in conjunction with tillage or at a different period during the growing season, is especially common in the production of many important crops. Such appropriate use of multiple methods is referred to as integrated weed control.

The advantages of managerial practices, such as planting a crop as early as possible to give it a competitive edge over weeds that appear later, are well known. But that procedure alone does not provide enough weed control to allow satisfactory crop yields; other procedures are essential. Mechanical control, or cultivation, is usually required, often in combination with one or more herbicide treatments.

The primary objective of integrated weed control is to provide effective control as economically as possible without adverse effects to the crop. Integration is possible by applying the various available methods to each weed-control problem. Such application may be made on either a small scale or a large scale on agricultural lands, on aquatic or industrial sites, or wherever control of weeds is desirable. Although integrated weed control has been used traditionally (Upchurch, 1975), new advances apparently can and should be made. Such advances can come, considering all available control methods and all factors involved.

FIGURE   9.9

Above, a fence row infested by weeds. Below, the same fence row after weeds had been controlled by an herbicide. [Courtesy of J. F. Miller.]

Progress is also possible in the integration of weed-control procedures with methods of controlling other pests. Additional emphasis, for example, may be given to controlling a weed that not only competes with the crop, but also serves as an alternate host to a particular fungus, virus, or insect that severely affects that crop. Integration of all procedures that are required to control existing pests with minimum cost and maximum efficiency is the objective of integrated pest management.

LITERATURE CITED

Andersen, R. N. 1968. *Germination and Establishment of Weeds for Experimental Purposes.* Champaign, IL: Weed Science Society of America.

Ashton, F. M., and A. S. Crafts. 1973. *Mode of Action of Herbicides.* New York: Wiley.

Buchanan, G. A. 1975. "Weed losses too high." *Weeds Today,* 6(1):4.

Crafts, A. S. 1975. *Modern Weed Control.* Berkeley: University of California Press.

Green, M. B. 1973. "Are herbicides too expensive?" *Weeds Today,* 4(3):14–16.

Irving, G. W., Jr., and E. L. LeClerg. 1965. *Losses in Agriculture.* U. S. Dept. Agr. Handbook, no. 291. Washington, DC: U. S. Government Printing Office.

Klingman, G. C. 1975. "Broad-spectrum and narrow-spectrum herbicides: A need for both." In J. C. Street, ed., *Pesticide Selectivity,* pp. 1–90. New York: Marcel Dekker.

_____, and F. M. Ashton. 1975. *Weed Science.* New York: Wiley.

Martin, J. H., W. H. Leonard, and D. L. Stamp. 1976. *Principles of Field Crop Production,* 3rd ed. New York: Macmillan.

Muzik, T. J. 1970. *Weed Biology and Control.* New York: McGraw-Hill.

Reed, C. F., and Regina O. Hughes. 1977. *Economically Important Foreign Weeds: Potential Problems in the United States.* U. S. Dept. Agr. Handbook, no. 498. Washington, DC: U.S. Government Printing Office.

Shaw, W. C. 1974. "Total farm weed-control concepts for pest management." Invitational address, annual meeting of Weed Science Society of America, February 12–14, 1974, Las Vegas, NV.

Upchurch, R. P. 1975. "Biological, chemical, and other factors influencing strategies for selective weed control." In J. C. Street, ed., *Pesticide Selectivity,* pp. 11–20. New York: Marcel Dekker.

10

Vertebrate Plant Pests and Their Control

Jerome V. Shireman

Vertebrate pests and their effects on crop plants are treated in this chapter. A vertebrate pest, according to Howard (1967), "is any native or introduced, wild or feral, non-human species of vertebrate animal that is currently troublesome locally, or over a wide area, to one or more persons, either by being a health hazard or a general nuisance, or by destroying food, fiber, or natural resources." Pests such as house-flies, mosquitoes, biting flies, ticks, mice, rats, and worms can be eradicated at will with little public resentment, but the addition of birds and mammals, which have esthetic or wildlife values, to the list of pests is frequently met with public resentment.

The goal of the professional pest-control specialist is usually not to eradicate but to keep the population numbers below the economic threshold. Many animals on the vertebrate-pest list have actually in-creased because of human activities; others, because of low tolerance to wide-spectrum pesticides, have decreased in numbers. The latter group of animals is often found on endangered-species lists and must be protected, whereas troublesome pests must be controlled. This is an awesome task, because the life histories of many vertebrate species are similar. Bird species, for example, cannot be isolated from one another. The widespread application of poisonous chemicals to cereal seeds could reduce the population of desirable songbirds and pro-tected wildlife species as well as the target bird species.

Vertebrate pests cause much damage to agricultural crops. Their effects are often localized and might be caused only by a segment of the population. The Quelea bird *(Quelea quelea)* in Africa provides a prime example of the damage that can be done to wheat crops; each bird consumes between two and three grams of food per day, but destroys at least eight times that amount as it moves through the field. Large flocks migrating from one cultivated area to another devastate entire regions. The damage they cause is estimated in excess of $200 million annually.

Most attention is directed to a vertebrate pest when it is in some way involved in the transmission of human diseases, and the control procedures currently used have resulted because of the importance of these animals as vectors of parasites. Mammals identified as vectors of the plague, for example, are often agricultural pests as well.

It is difficult to assess the role of certain birds and mammals as plant pests. Often the animal identified as destructive by some is not considered a pest by others. Many of the birds and mammals considered pests by agriculturalists are considered desirable by sportsmen and conservationists, and such wildlife may be protected by state and federal regulations. Other animals, such as the muskrat, nutria, and beaver, are valued for their fur. These animals, however, may inflict considerable damage to crop plants. Control of these valued animals is often questioned by special-interest groups whose livelihood does not depend on crop plants.

METHODS OF VERTEBRATE-PEST CONTROL

Seven categories of plant-pest-control methods were listed in Chapter 5: exclusion, eradication, therapy, vertical resistance, horizontal resistance, protection, and avoidance. Several of these methods (exclusion, eradication, protection, and avoidance) apply readily to vertebrate-pest control, but therapy, vertical resistance, and horizontal resistance do not apply directly.

During the evolution of pest control, several general types of control methods have evolved. These include chemical, biological, and mechanical methods. All are used to control vermin, and, depending on how they are applied, can be used for eradication and exclusion of vertebrate pests and for protection of crop plants.

Chemical Methods

Chemicals are used not only to eradicate or control vermin, but also as repelling and frightening agents. Several of the chemicals act as sterilants and will be discussed under biological control methods.

The diverse chemicals used as rodenticides are toxic to all animals (Peoples, 1970) and can generally be grouped into organic compounds, inorganic compounds, and organic fumigants. Of the organic compounds, the most popular are those that inhibit the coagulation of blood. These compounds interfere with the conversion of prothrombin to thrombin in the liver, and thus cause hemorrhages throughout the body by rupturing the capillaries. They eventually cause the animal to bleed to death internally. There is another group of organic compounds that act by entering metabolic pathways (citric-acid cycle) and are extremely toxic to all animals. Included in this group are the notorious sodium fluoroactate, commonly known as 1080, and strychnine sulfate, an alkaloid, which causes animals to go into convulsions.

Of the inorganic compounds, some of the more common ones in use are zinc phosphide, thallium sulfate, barium carbonate, yellow phosphorus, and white arsenic. Several fumigants are also used. These include carbon monoxide (combines with hemoglobin), methyl bromide (causes pulmonary congestion), and hydrogen cyanide (inactivates cytochrome oxidases). None of the compounds mentioned above are specific to any group of animals, and all are dangerous to humans. Because they are nonspecific, they must be used with precision to avoid killing nontarget species. Because evidence exists that many target animals are developing genetic resistance to the widely used anticoagulant compounds (Gratz, 1973; Jackson and Kaukeinen, 1972), emphasis is being placed on the use of target-specific rodenticides and avicides.

Chemicals are also used for frightening birds and mammals; these can be divided into lethal and nonlethal groups. Nonlethal chemicals are the most desirable, because accidental poisoning of upland game animals is avoided. The chemical agents phenyl N-methyl carbamates (DRC-736) and 4-amino-pyridine (DRC-1327) have shown promise for bird control. DRC-1327 was used in feedlots to discourage redwing blackbirds. Cracked corn was treated with the chemical and broadcast at one-month intervals throughout the feedlot. Redwing populations fell sharply, with less than 20 percent mortality (Woronecki et al., 1967). Protection was offered to cornfields by spraying DRC-1327 on partially husked ears and broadcasting cracked corn in the fields. The number of birds feeding in the treated field decreased, and few nontarget species were killed.

Other uses for chemical repellents are the treatment of conifer seeds against both rodents and birds, and the treatment of seedlings and trees against destruction by larger mammals. Increased needs for reforestation have placed greater emphasis on protecting pine seeds from small mammals. Endrin has been used for this purpose with success; however, this chemical can be classified as a lethal agent, and for this reason other, nontoxic chemicals would be more desirable. A chemical that has shown promise is mestranol, a methylated derivative of natural estrogen. Trial tests using this chemical on Douglas fir seed have met with some success (Lindsey et al., 1974).

Chemical repellents applied to the bark of dormant trees, conifer seedlings, and orchard trees have been successful for repelling rabbits, deer, and porcupines. Several chemicals, trinitrobenzene-aniline (TNBA), zinc dimethyldithiocarbamate cyclohexylamine (ZAC), and tetramethyl thiuram disulphide (TMTD), have broad application for this purpose (Welch, 1967).

Biological Control Methods

Biological control is the use of a biological agent, not necessarily a vertebrate species, against the pest organism to create control. The most common biological control procedure is the introduction of a predator into the pest population. The predator must be studied carefully before introduction to ensure that it will not become a pest, because predators are not obligate and often prey on species other than the target species (McCabe, 1966), especially when placed in new situations. Moreover, animal populations are usually not regulated solely by predators (McCabe, 1966), and it is doubtful that these populations can be kept under control only by introduction of a predator species. In fact, predation pressure might actually increase the density of the prey species by cropping animals that are weak and unfit for reproduction (Howard, 1967). The animals that remain are the reproductively active individuals. Even when predators do temporarily reduce a prey population, it is doubtful that the population is reduced enough to establish control (Howard, 1967).

To introduce pathogenic organisms is another way to control populations biologically. When introduction of such organisms into pest populations is contemplated, a thorough study should be conducted to identify the potential hazards of the introduction, since the organisms are usually exotic rather than endemic. If an endemic organism has not previously suppressed the pest population, it is not feasible to reintroduce that organism. According to Herman (1964), the following factors must be considered before disease organisms are introduced: (1) the biological-control organism must be highly pathogenic to the target species; (2) the killing power of the organism must be anticipated; (3) the organism must be host-specific; (4) the organism must be available and the environment must be suitable for its perpetuation and spread; and (5) the control program should be monitored to establish the actual adverse and beneficial effects of the introduction.

Possibly the classic example of using a pathogenic organism for vertebrate-pest control was the introduction of a virus (the agent of myxomatosis) into Australia for control of the European rabbit *(Oryctolagus cuniculus)*. The results of this introduction were dramatic, and the resulting epidemic caused a considerable decline in the rabbit population (Fenessy and Mykytowycz, 1974). Mosquitoes are the principal vectors of the virus. Rabbit populations were not annihilated by the disease; instead, the rabbits developed genetic immunity (Tomlinson and Gooding, 1970). Nevertheless, populations of rabbits were reduced to a point where other control methods can now be used to keep them in check.

Reproductive Control Methods

The concept of sterility created by use of chemosterilants, radiation, or genetically sterile individuals was given a tremendous boost with the development of radiation-induced sterility in male screwworm flies *(Cochliomyia hominivorax)*. Research on chemosterilants has been encouraged in recent years by pharmaceutical companies. Most of their research, however, has been directed toward the physiological effects of drugs that may be used as human contraceptives. The results obtained in these studies have proved useful for controlling rodent pests because laboratory mice were used as the experimental animals. Thus, potential chemicals for wild-animal control were identified. Although studies have indicated that chemosterilants offer possibilities for vertebrate-pest control, the technique is not is general use. Of great importance, from the environmentalist's viewpoint, is that chemosterilants lessen the need for lethal chemicals, thus reducing the threat of environmental contamination. This approach to regulating the density of a pest population shifts the emphasis away from increased mortality, and places it on decreased natality (Marsh and Howard, 1970).

Habitat Management

Another biological pest-control method is habitat alteration. Areas can be made either unsuitable for habitation or more suitable, and thus pests can be repelled from crop areas or attracted to areas not under cultivation. Wildlife refuges and sanctuaries can be used for this purpose. An animal is found within a particular area because the necessary ingredients for its survival are present there. An undesirable consequence of habitat manipulation, however, is the possibility of altering the entire ecosystem (Howard, 1967).

Usually an animal becomes a pest because it becomes adapted to a particular habitat created by human beings. Blackbirds, grackles, and starlings often give up their foraging for wild foodstuffs when cattle feedlots and grain fields are in the area. They are attracted to these banquet sites, where they consume large quantities of grain and spoil additional amounts by fouling food containers and stored grain. Waterfowl often find cornfields and ricefields more inviting than their natural feeding grounds, and large concentrations of these birds cause serious damage to these crops. Other birds such as robins and bluejays forsake wild fruits, which are harder to find, and seek vineyards, berryfields, and orchards, where food is plentiful.

Often vertebrate-pest problems can be reduced by decreasing suitable habitats near farming operations. Clean-farming practices, such as removing cover along fence rows and woodlots near orchards, and using field storage of grain, reduce pest problems, but may also destroy the habitats of nonpest species.

Another method whereby depredation of valuable crops might be decreased is the planting of more inviting food crops in the area. White-tailed deer, for example, will often feed on young conifer trees. If more suitable browse plants are provided, deer will feed on these and ignore the young pine trees.

Vegetation barriers can also be used to keep pests away from valuable crops. The results obtained from trial plantings, however, seem conflicting. Lewis (1946) reported vegetative barriers to be effective in protecting crops from jackrabbits, whereas Evans et al. (1970) reported just the opposite. These authors further point out that habitat alteration exclusively for jackrabbit control may not only have little effect on alleviating damage due to jackrabbits, but may also conflict with other land-use projects.

The establishment of specific land areas for wildlife management can be used for animal control if these areas are attractive to wildlife species. Waterfowl depredation on rice and other crops can be reduced if suitable feeding and nesting areas are provided. Waterfowl can often be frightened from crop lands to nearby refuges, where they can be held until crops are harvested.

Surfactants

Surfactants are surface-active agents that lower the surface tension of water. Application of these agents to birds kills them. Although the exact physiological mechanisms are not known, wet plumage combined with rainfall and cold temperatures apparently causes body temperatures to drop to lethal levels. These wetting agents are usually applied over roosting areas, where many birds can be "wetted" at one time. If the proper climatic conditions prevail after treatment, considerable mortality can be expected.

Mechanical Methods

Mechanical devices can be grouped according to the function of the device, and have probably been used more than any other control method.

Traps

According to Fitzwater (1970), trapping is the oldest profession known to man. Traps were originally designed to capture animals for food and fiber rather than to control their numbers. Traps vary in design and construction, and have been designed to capture all kinds of vertebrates. They can be very simple, improvised traps, such as dead falls, or sophisticated traps that combine lights and gas chambers, where animals can be killed upon capture. Large baited-cage traps are probably the most efficient kinds for catching large numbers of animals, and are used with considerable success to control bird populations.

Frightening Devices

Various means have been employed to frighten pests away from crops, and usually work best for bird pests. Frightening devices can be grouped according to their means of control. Exploders have been used on roosting concentrations to kill birds and to scare them away. Gas and shrapnel bombs are efficient in roost areas and are selective to the species using the roost; Neff and Meanly (1957) reported that 300,000 redwings, grackles, cowbirds, and starlings were killed with bombs in two consecutive seasons. They found only one dead bird of another species. Roost bombing is hazardous and should be performed only by trained personnel. Automatic gas exploders, guns, and firecrackers are also used to frighten birds from fields. Gas exploders are usually placed to project the sound over the field. For maximum efficiency, they should be moved frequently and used with other frightening devices.

Loudspeakers and sirens are used to scare birds from concentration areas. These devices can be mounted on the ground or in airplanes. Although birds are frightened and made to fly, they often move only from one location in the field to another. Another disadvantage to using sounds of this type is that birds become accustomed to them if they are applied for long. The time for sound conditioning depends on the species; some may condition to the sound quickly (redwing blackbirds, cowbirds), whereas other habituate more slowly (starlings, yellowheaded blackbirds). Another method that uses amplified sound is to broadcast alarm sounds and distress calls. Results with this method have been varied, but success probably depends on techniques. It is best to direct the proper alarm sound to the proper species. As an example, blackbirds will not respond to starling sounds, but starlings will respond to blackbird sounds; cowbirds are immune to most sounds. Boudreau (1967) reported that a well-executed program using alarm calls decreases the target population by as much as 95 percent for a period of four or five days.

PRINCIPLES OF CONTROL OF AVIAN AND MAMMALIAN PESTS OF PLANTS

Control of birds and mammals will now be treated according to the principles discussed in Chapter 5, although, as has already been mentioned, some of these principles are not applicable to vertebrate pests. Only a few plant varieties have been developed that show either vertical or horizontal resistance to vertebrate pests, but several varieties of grain sorghum are resistant to blackbirds. These could be used as crop substitutes for corn where depredation occurs. The resistance of sorghum is believed to depend on the brown color of its seed. The brown coloring may result from astringent chemicals that repel the birds. Corn varieties that have drooping ears suffer less blackbird damage than varieties with upright ears.

Therapeutic treatments made after the plants has been attacked by its pest are of little value for controlling vertebrate pests. Vertebrate pests, unlike other pest organisms, leave the damaged plant almost immediately after attack. Therapeutic methods, therefore, would offer control only if the treatment repelled the attacking organism upon later visits, whereupon the treatment might be classified as protective.

Birds

Certain birds cause serious problems in agricultural areas because of their tendency to congregate in large flocks, which then consume large quantities of food material. They damage plants during all stages of the plants' life cycles. The most serious depredations occur in grain and fruit crops as the fruit or grain nears maturity. Such birds as finches, geese, and pheasants do considerable damage after seed germination. Geese, for example, graze young plant shoots; yet, for this reason, domestic geese have been used to weed some agricultural crops, such as sugar beets, upon which geese do not graze.

Exclusion

It is generally difficult to exclude birds from an agricultural area. Exclusionary methods are successful only before an exotic bird has been introduced into a country or geographical area. The importation of alien animals into the United States is supervised by the Bureau of Sport Fisheries and Wildlife. Their functions are to ensure that no prohibited species enters the United States, to enforce compliance with Federal Bird Acts, and to assist states in enforcement of state laws and regulations concerned with entry of wildlife species that are not under federal prohibition. Since 1900, more than twenty Federal Acts for the protection of wildlife have been passed. Four of these are especially important for bird control: The Migratory Bird Act of 1918, as amended (most birds found in North America are classified as migratory); the Lacey Act of 1900, as amended, which regulates what kinds of wildlife can be imported; the Tariff Act of 1940, as amended, which imposes import duties on waterfowl and regulates importation of certain bird feathers and skins; and the Bald Eagle Act of 1970, as amended, which not only protects our national bird, but restricts the importation of Bald and Golden Eagles from any foreign country.

Prior to 1966, importers were not required to report the number of live animals brought into the United States. Importation figures for 1969 show that approximately 573,000 wild birds, 116,000 mammals, 74,000,000 fish, 339,000 amphibians, and 1,394,000 reptiles were imported. The task of inspection and identification of these organisms becomes extremely burdensome. Permits must be obtained to import all birds except four species, whereas certain other birds cannot be imported. The list of banned species includes the Java sparrow, *Padda oryzivora;* the Sudan dioch, *Quelea quelea;* the rosy pastor, *Sturnus roseus;* and the red-whiskered bulbul, *Pycnonotus jococus.*

Eradication

Because of public opinion, federal laws, and widespread distribution and mobility of birds, it is virtually impossible to eradicate bird pests in the literal sense. For these reasons, bird populations are kept in check, often by eradicative procedures, in agricultural areas when they become troublesome. Chemicals are frequently used for eradicative purposes, but must be used with caution, since most chemicals are not selective. Initially, highly toxic materials like 1080, strychnine, and thallium sulfate were applied to baits that were presumed specific to the pest species. Toxic chemicals were also applied to artificial perches. More recent research has been directed toward the development of species-selective chemicals such as avitrol (4-amino pyridine) and starlicide (3-chloro-p-toluidine hydrochloride). These chemicals are directed against a specified group of birds. Other avicides, specific to problem birds but having other modes of action, include both chemosterilants and surfactants.

Before adequate control can be achieved, the behavior of the selected bird pest must be known. This is important for two reasons: first, the control must be achieved without harming nontarget species; and, second, the control must be applied at the proper time to achieve maximum results.

Several bird species referred to as "blackbirds" (redwing blackbird, starling, yellow-headed blackbird, common grackle, brown-headed cowbird, and Brewer's blackbird) are serious crop pests. The redwing blackbird is most abundant and economically the most important in terms of depredations of agricultural crops (Boudreau, 1967). In the southeastern United States, redwing blackbirds, grackles, cowbirds, and starlings may be found in the same roost area. In the Southwest, redwing blackbirds, yellow-headed blackbirds, Brewer's blackbirds, starlings, and cowbirds are found in associations during the winter. Blackbird roosts are usually found in brush, hardwoods, and conifers in upland situations, and in marsh vegetation, usually over or near water. Grackles, cowbirds, and starlings are usually present in these roost situations. Dense overhead vegetation and proximity to the water probably give good protection against both avian and terrestrial predators in wetland roosts.

Birds begin moving to the roost area before sunset and will often preroost near the night-roosting site. Usually preroosting birds are well-fed birds returning early from feeding areas. Late-returning birds will immediately enter the roost area. Birds leave the roost just before sunrise and fly directly to feeding areas. Grackles usually leave the roost before other blackbirds.

In the spring, these large concentrations break up, and breeding takes place. Food habits also change during the spring, and even though flocks may be seen feeding in fields, they feed mostly on animal material. Blackbirds damage ripening grain crops, stored grain, and grain in feedlots. Most of the damage occurs during the fall and winter, when the birds are found in large concentrations. Less damage occurs in the spring, when germinating sprouts are pulled. The logical time for reducing populations is during high depredation periods, when blackbirds are gathered in flocks around the fields and in roosts. The methods most often used to control blackbird populations include trapping, poisoned baits, roost bombing, and roost spraying.

Trapping.   Trapping success varies with the type of trap used, season, bird concentrations, trap placement, food availability, and trapper experience. Large decoy traps are usually used for blackbirds, and are highly selective. The trap is made of poultry wire and kept stocked with decoy birds, food, and water. Any species other than blackbirds collected can be released unharmed. The disadvantage, however, is low returns for the number of man-hours spent. Floodlight traps are more successful. These are large, funnel-shaped structures of netting that taper back to a tent containing bright lights. This trap is operated at night after the birds have returned to their roosts and have settled down for the night. The trap works best in roosts located in low, dense vegetation, where the trap mouth can be placed within a few feet of the roosting birds. After the trap is set and the floodlights lit, the roost is encircled by several men and the birds are herded toward the trap. After capture, the birds are held within the closed tent where they can be quickly killed with lethal gas. This trap is very efficient if conditions are ideal. As many as 120,000 blackbirds and starlings have been captured in a single night (Meanley, 1971).

Poisoned Baits.   Unless poisoned baits are used properly, they can be dangerous and nonselective to other bird species. Before baits are used, the habits of the unwanted birds should be studied. Knowledge of their feeding habits and breeding areas will help to establish proper bait placement. To ensure that the baited material is ingested, proper bait materials must be used.

Roost Bombing.   Bombs using explosive charges filled with lead shot have been used to kill birds in roost areas. This method has lost favor, however, because many birds are crippled, and because it is time-consuming and dangerous to the operator.

Roost Spraying.   Initially, roosts were sprayed with lethal contact poisons delivered by airplane. The chemicals used were concentrated and dangerous to humans, wildlife, and livestock. More recent research has been aimed toward developing materials that are lethal to birds but not to other organisms. Surfactants meet this requirement. Birds are sprayed with a wetting agent, and if proper weather conditions prevail, many birds are killed.

Nonroosting Birds.   Birds such as robins, Baltimore orioles, rose-breasted grosbeaks, and cedar waxwings become serious pests in orchards, vineyards, and berryfields. Control of these birds is especially difficult because of federal and state laws and public sentiment against killing songbirds. Whenever possible, these birds are frightened from fields, or the crop is protected in some manner. Most eradicative methods are impractical, not only for the above reasons, but also because these birds are often widely dispersed and when eradicated are replaced by other birds in the vicinity. In areas where these birds cause serious losses to crops, poisoned baits are used with some success.

Another area where birds become serious pests is in and around food-processing plants. Processed foods are contaminated directly with feathers, droppings, regurgitated pellets, seeds, and ectoparasites. Birds also do damage directly by eating stored food products. Indirectly, they may cause the spread of disease. Several species of *Salmonella*, as well as *Histoplasmosis* and *Crytococcus*, are spread by bird droppings. These organisms grow in accumulated dung (Keown, 1973).

Eradicative methods used in these areas include shooting, fumigants, toxicants (wetting agents, toxic perches, contact poisons, and ingested materials), and trapping. In one situation where cedar waxwings were causing contamination to raisins, the birds were trapped and removed from the area. They were transported twenty miles from the site and released unharmed (Palmer, 1973), with no adverse effects to the birds or complaints from local inhabitants.

Protection

As defined in Chapter 5, protective measures are those in which a barrier—either chemical or physical—is placed on or immediately around growing plants or plant products to prevent pests in the vicinity from establishing pest-victim relations with them. Protective measures are used in vertebrate-pest control, and may consist of both chemical and physical barriers. Mechanical frightening devices, chemical frighteners, clean-farming practices (removal of pest sites, etc.), and enclosures are used to protect both field and stored crops. These methods are employed whenever possible, because they both protect the plant and decrease the possibility of harming valuable nontarget species.

Because game and valuable songbirds often cause serious crop depredations, the trend has been to develop and use nontoxic chemical repellents. Some chemical repellents, avitrol, for example, are toxic to birds and are used in feedlots and other areas frequented by blackbirds. A small percentage of the flock may be killed, but the affected birds become disoriented and cause the entire flock to leave the area. Other chemicals such as DRC-3324 (abutyramide) are being investigated as nonlethal frightening agents. Laboratory studies indicate that this chemical acts as an effective repellent well below toxic levels (Frank et al., 1970). In field tests, DRC-3324 was an effective repellent against common grackles and redwing blackbirds. Besser et al. (1973) presented convincing evidence for the effectiveness of chemical frightening agents for protecting crops. Data were collected from 1962 through 1971 at the Sand Lake National Wildlife Refuge, South Dakota. In 1962, blackbirds roosting on the refuge fed in surrounding agricultural lands. The bird population was estimated at 1,345,000 birds in 1962. Starting in 1964, avitrol treatments were started. During the seven years of treatment, the blackbird population declined to 250,000 birds. Tag returns indicated that the decline in roosting birds did not result from a decline in bird populations on the breeding grounds but from the fact that the birds learned to avoid the treated area and changed migration routes.

Protective barriers for bird control are usually impractical. But, in certain situations, such as in vineyards and berryfields, where the crop is valuable and does not cover an extensive area, these methods may be practical. Initially, protective netting is costly in both materials and labor. There are, however, several advantages of this control method: it is effective (nearly 100 percent) and nontoxic, and the nets are reusable. Plastic netting, for example, can be reused many times if stored properly.

Stored and processed foods can be adequately protected from depredations by physical means. Storage bins made of screen and sheet metal are commonly used for this purpose. Architectural design should be considered in the construction of processing plants. Nesting and perching places (ledges, etc.) should be eliminated. Generally, birds can be kept out of the plant if screens are employed and proper maintenance of the facility is practiced.

Avoidance

Serious damage to crops can be curtailed if major concentrations of pests are avoided. Crops can be planted or varieties used that will ripen at periods when pests are not concentrated. In areas where crops cannot be planted to coincide with low pest numbers, it is also important to have all fields ripen at nearly the same time. In the Arkansas rice belt, for example, it is impossible to avoid depredation by adjusting planting and harvesting dates. In this situation, early and late maturing fields receive serious depredation, because birds concentrate in these fields. But if plants in all fields ripen at the same time, the chance is that the flocks will disperse over a greater area, doing less damage to isolated fields. The same factors apply in the corn belt; usually, fields that mature out of phase are severely damaged. De Grazio (1964) reported that a tight-husked hybrid corn variety was evaluated for bird resistance. This was a slow-maturing variety and was in the dough stage when the corn on plants of the surrounding fields was beginning to harden. This field, because it was more suitable to the birds, suffered greater damage than surrounding fields. In Louisiana and Texas, farmers can plant early-maturing rice varieties. These varieties ripen in late June and July, before blackbirds begin forming flocks (Meanley, 1971).

Mammals

Exclusion by means of legal regulation has been discussed in the previous section. Generally, most laws and regulations that pertain to birds also pertain to the exclusion of mammalian species from the United States. Some mammals excluded from importation are: the mongoose, including any species of the genera *Atilax, Cynictis, Helogale, Herpestes, Ichneumia, Mungos,* and *Suricata;* bats of the genus *Pteropus;* multimammate rats of the genus *Mastomys;* the East Indian wild dog of the genus *Cvon.* As with birds, plant species with vertical and horizontal resistance have not been developed for mammalian pests. Eradicative and protective methods, however, have widespread use in mammalian control.

Eradication

It would be extremely difficult and expensive to eradicate unwanted mammal populations from agricultural regions. Pest mammals include a wide variety of animals and, therefore, diverse control methods are needed. The best approach is to maintain a population level that is tolerable to the farmer, storage operator, and processor.

Small mammals belonging to the class Rodentia are especially damaging to orchards, forest plantings, and stored plant products. These small mammals damage plants during all stages of the plant life cycle. They eat seeds and sprouting plants, damage root systems, girdle trees, and consume and contaminate stored crops.

The most common control method is to use toxic chemicals, such as strychnine, 1080, and zinc phosphides. They are usually applied to bait materials with no guarantee that nontarget species will not be killed. Treated baits can be broadcast, placed in special bait containers, or placed in artificially dug burrows. Ground sprays are also used, especially in orchards, in order to control pine mice. In the past, endrin was used for this purpose, but its use is currently under federal restriction.

Anticoagulant chemicals, such as warfarin, pival, and fumarin, are commonly used because they are essentially target-specific. Evidence suggests, however, that natural resistance to these chemicals is being developed by certain species. Another anticoagulant that has shown promise is chlorophacinone. This chemical has been used experimentally as a ground spray (Horsfall, 1972) in apple orchards. Not only were adult animals nearly eliminated, but reproduction was reduced as well. This chemical has a short residual effect, and was not detected in either the fruit or the runoff water from experimental plots. The effectiveness of ground spraying, however, was related to the amount of food available, because the poison had been conveyed to the animals through their food supply. The most common way to use chlorophacinone is to mix the poison with bait material (wheat, oat, groats, etc.) and either broadcast the mixture or place it in bait stations.

The best approach to chemical-control practices is to time them to coincide with the damaging activities of the animal. Treatment should be coordinated with the behavior of the target species. As an example, pine mice are a serious problem in orchards in the northeastern United States. Meadow mice are also present there, but the damage caused by them is not nearly as serious. In this situation, placement of the bait is important. Broadcast baits are eaten more readily by meadow mice, whereas baits placed in burrow systems more effectively reduce the number of pine mice.

Reproductive inhibitors are also being investigated as possible population-control methods. The advantages of this method are: they do not create hazards to other wildlife; they cause less pollution of the environment; and they could be used to maintain populations at desired levels. Most studies to date have been laboratory work, intended to establish the physiological effects of various chemicals and to identify chemicals that might have field application.

Mestranol (17-ethynyl-3-methoxyestra-1,3,5(10)Otrien-17-ol) is a steroid hormone that has been effective when given to rodents. This hormone was field-tested on Richardson's ground squirrels (Goulet and Sadleir, 1974). Females given the drug before or during early pregnancy were sterilized, and birth rates were reduced. Reduction in aggressiveness and metabolic rates were noted in treated females.

The effects of the treatment, however, were nullified during the first year because of increased survival of young animals and immigration of animals from surrounding areas. Results from the second season were more encouraging. Reduced population numbers on the control plots were not compensated for by immigration, and the population crashed. From these results, it appears that chemosterilants show promise for controlling certain wild-animal populations and should be relied upon more frequently in the future as a control method.

Trapping is often used in conjunction with other control methods in order to discover both initial population numbers and the effectiveness of treatments. This method has limited application for controlling most small mammals. Trapping is used to control mammals such as beavers, moles, nutria, muskrats, and armadillos. These animals live in burrows, dens, or lodges, and are susceptible to trapping because of their restricted movements within burrows or runs. Either "live" or steel traps are used, and the animal is trapped as it moves from one area to another.

Protection

Protective methods discussed in the bird section apply also to mammals. Some methods, such as fencing, can be used more effectively against mammals, such as deer and rabbits, than against birds, since mammals cannot, of course, fly over the fences. Individual protective measures are also practiced. Trees, for example, can be protected from rabbits and porcupines with aluminum or nylon strips. The animals are discouraged from chewing through the barrier to girdle the tree. Installation of these barriers throughout wide areas is impractical, however.

Crops might also be protected from mammal damage by the planting of fringe crops. If the fringe planting is preferred by the animal, it will consume this plant rather than the harvestable crop. Fringe crops have been used to discourage jackrabbits from feeding in certain fields. These methods offer some reprieve from damage when population numbers are low, but are usually not successful when they are high (Evans et al., 1970). Land barriers have also been used, but the results are erratic.

Chemical repellents seem to offer promise for plant protection. The chemical must not harm the plant and must be distasteful to the pest. Several chemicals have been tested for this purpose and have proved successful in certain situations. These chemicals, however, have not yet been approved for widespread application.

LITERATURE CITED

Besser, J. F., J. W. De Grazio, and J. L. Guarino. 1973. "Decline of blackbird population during seven years of baiting with a chemical frightening agent." *Proc. Sixth Bird-Control Seminar,* Bowling Green, OH, pp. 12–13.

Boudreau, G. W. 1967. "Blackbird behavior." *Proc. Third Vert. Pest-Control Conf.,* San Francisco, CA, pp. 57-60.

De Grazio, J. W. 1964. "Methods of controlling blackbird damage to field corn in South Dakota." *Proc. Second Vert. Pest-Control Conf.,* Anaheim, CA, pp. 43-49.

Evans, J., P. L. Hagdel, and R. E. Griffith, Jr. 1970. "Methods of controlling jackrabbits." *Proc. Fourth Vert. Pest-Control Conf.,* West Sacramento, CA, pp. 109-116.

Fennessy, B. V., and R. Mykytowycz. 1974. "Rabbit behaviour research in Australia and its relevance in control operations." *Proc. Sixth Vert. Pest-Control Conf.,* Anaheim, CA, pp. 184-187.

Fitzwater, W. D., 1970. "Trapping: the oldest profession." *Proc. Fourth Vert. Pest-Control Conf.,* West Sacramento, CA, pp. 101-108.

Goulet, L. A., and R. M. F. S. Sadleir. 1974. "The effects of a chemosterilant (Mestranol) on population and behavior in the Richardson's ground squirrel *(Spermophilus richardsonii)* in Alberta." *Proc. Sixth Vert. Pest-Control Conf.,* Anaheim, CA, pp. 90-100.

Gratz, N. G. 1973. "A critical review of currently used single-dose rodenticides." *Bull. Health Org.,* 48:469.

Herman, C. H. 1964. "Disease as a factor in bird control." *Proc. Second Bird-Control Seminar,* Bowling Green, OH, pp. 112-121.

Horsfall, F., Jr. 1972. "Chlorophacinone and herbage as potentials for pine mouse damage control." *Proc. N. Y. Pine-Mouse Symposium,* Kingston, NY, pp. 32-46.

Howard, W. E. 1967. "Biological control of vertebrate pests." *Proc. Third Vert. Pest-Control Conf.,* San Francisco, CA, pp. 137-157.

Jackson, W. B., and D. Kaukeinen, 1972. "Resistance of wild Norway rats to anticoagulant rodenticides confirmed in the United States." *Pest Control,* September, 13–14.

Keown, J. W. 1973. "Pest-bird control in and around food-processing plants." *Proc. Sixth Bird-Control Seminar,* Bowling Green, OH, pp. 17-18.

Lewis, J. H. 1946. "Planting practice to reduce crop damage by jackrabbits." *J. Wildl. Mgmt.,* 10:227.

Lindsey, G. D., R. M. Anthony, and J. Evans. 1974. "Mestranol as a repellent to protect Douglas-fir seed from deer mice." *Proc. Sixth Vert. Pest-Control Conf.,* Anaheim, CA, pp. 272-279.

Marsh, R. E., and W. E. Howard. 1970. "Olfaction in rodent control." *Proc. Fourth Vert. Pest-Control Conf.*, West Sacramento, CA, pp. 64-70.

McCabe, R. A. 1966. "Vertebrates as pests: a point of view." In *Scientific Aspects of Pest Control*, pp. 115-134. Washington, DC: Nat. Acad. Sci. Nat. Res. Council.

Meanley, B. 1971. *Blackbirds and the Southern Rice Crop.* Washington, DC: US Dept. Interior, Fish and Wildl. Ser. Res. Publ., no. 100.

Neff, J. A., and B. Meanley. 1957. "Research on bird repellents, progress report no. 2: bird-repellent studies in the eastern Arkansas rice fields." Denver, CO: U.S.D.I. Bur. Sport Fish. and Wildl., Denver Wildl. Res. Lab. (mimeo).

Palmer, T. K. 1973, "Bird management on the farm and at the processing plant." *Proc. Sixth Bird-Control Seminar*, Bowling Green, OH, pp. 15-16.

Peoples, S. A. 1970. "The pharmacology of rodenticides." *Proc. Fourth Vert. Pest-Control Conf.*, West Sacramento, CA, pp. 15-18.

Schafer, E. W., Jr., and J. L. Guarino. 1970. "Problems in developing new chemicals for bird control." *Proc. Fifth Bird-Control Seminar*, Bowling Green, OH, pp. 7-10.

Tomlinson, A. R., and C. D. Gooding. 1970. "The organization of rabbit control *(Oryctolagus cuniculus)* in Western Australia." *Proc. Fourth Vert. Pest-Control Conf.*, West Sacramento, CA, pp. 188-194.

Welch, J. F. 1967. "Review of animal repellents." *Proc. Third Vert. Pest-Control Conf.*, San Francisco, CA, pp. 36-40.

Woronecki, P. P., J. L. Guarino, and J. W. De Grazio. 1967. "Blackbird-damage control with chemical frightening agents." *Proc. Third Vert. Pest-Control Conf.*, San Francisco, CA, pp. 54-56.

III
SYSTEMS OF PLANT-PEST CONTROL

11

The Systems Approach
to Plant-Pest Control

The science of plant-pest control (Part I), as practiced by controlling
the various kinds of plant pests (Part II), achieves full stature only if
we integrate our knowledge, now scattered in four related but distinct
disciplines: plant pathology (including agricultural nematology), entomology, weed science, and wildlife management. Important aspects
of each discipline deal directly with crop plants and the production,
transport, storage, and marketing of plant products. Plant pests abound
among the species of living things that each discipline treats, and
these pests have one characteristic in common: they all reduce the
quantity or quality of that which is produced by their crop-plant victims. They all behave as retarding factors. And all plant pests act and
react toward one another as well as toward their crop-plant victims,
whose products they destroy, often competing like jackals among
themselves over the remains.

The complex interactions among crop plants, their pests, and the
surrounding environment can be visualized in a general and much
simplified way as a group of overlapping circles (Figure 11.1). Depending on the circumstances, a given pest may react only with its
environment. Or, upon reaching its victim, it may interact with its
victim and its environment, which affects not only the pest and the
victim but also the interactions between the two. Finally, these already complex interactions may be further compounded by interac-

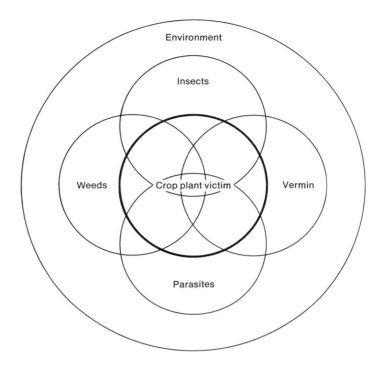

FIGURE   11.1

Interrelations between crop-plant victims, their pests (insects, weeds, vermin, and parasites), and the environment.

tions of all the original components with one, two, or three additional kinds of pests.

The science and technology of modern agriculture have now advanced to the point that plant-pest control can contribute its full share to the production of food and fiber only if it unifies its knowledge, scattered in four disciplines and, thus far in this book, in ten chapters. The only way to integrate our knowledge of plant-pest control into such a unit as depicted in Figure 11.1 is through the systems approach.

THE NATURE OF SYSTEMS

"When one is considering systems, it is always wise to raise questions about the most obvious and simple assumptions" (Churchman, 1968). Everyone surely has a concept of systems, a concept arrived at intuitively. Every schoolchild knows of the solar system; every person is aware of his or her own digestive system, circulatory system, and respiratory system; and all adults, it seems, consider themselves experts on religious and political systems. What are the fundamental elements of these and other such diverse systems? First, they all seek to make something work; each performs its own function. Second, they all have characteristic forms; each possesses a structure that makes possible the proper performance of its own function.

But a system is not just a single object having form and performing a function. System is a collective noun, because a system consists of several objects or units, all functioning in concert, like the musicians of a symphony orchestra. For example, the circulatory system of crop plants functions to transport water and nutrients and to translocate elaborated food materials. These functions, as we have seen in Chapter 3, are performed by specialized units within the system: xylem vessels may carry water and minerals in a kind of conduit; sieve tubes of the phloem mysteriously transport sugars and other substances bidirectionally throughout the plant; and so on. And even as the circulatory system consists of smaller functioning components, it serves as a part of a larger system, the crop plant itself. In the same way, our solar system, with its planets and satellites, is part of a larger system, a galaxy, and the galaxy is only a part of a still larger system; finally, the entire collection of systems within systems becomes the cosmos.

A third characteristic of a system is its interdependence with related systems. Systems do not perform in a vacuum. To cite again the circulatory system of a crop plant: water and nutrients will not be delivered to it for transport unless the root system has also performed its functions of absorption and accumulation; no elaborated food materials will be translocated through the phloem unless they have been manufactured earlier by the photosynthetic apparatus. The interdependence among systems is an essential element in our definition or concept of system.

A Definition of Systems

To reduce the complexities of even a simple system to a concise definition is difficult. True, a system is "a limited section of the real world" (Zadoks, 1971), but this brief characterization was not intended to be a definition. Watt (1966) defined a system as "an interlocking complex of processes characterized by many reciprocal cause-effect pathways," but I prefer to paraphrase the dictionary definition of a system: A system is a complex of ideas, principles, etc., arranged in a rational dependence or connection to form a coherent whole.

The etymology of the word gives the best notion of what system means. System is from the Greek, by way of Late Latin, and originally meant "to place together." The integration of ideas and principles is the essence of the systems point of view and of the systems approach to complex problems, such as those attendant on plant-pest control. In the vernacular, a system permits our "putting it all together."

Complex Systems

Although all systems are complex enough, some are certainly more complex than others. Watt (1970) arbitrarily defines a complex system as one having twenty or more variables. The complex system is thus difficult to comprehend or to mimic in the form of a working model. Its complexity shapes its nature.

An undesirable characteristic of complex systems is that they are invariably counterintuitive (Forrester, 1969). That is, complex systems suggest solutions that are ineffective, even wrong (Churchman, 1968). For example, the spruce budworm has been recognized as a significant pest for more than 200 years in the Canadian maritime province of New Brunswick, where it was devastating for periods of about eight years each in cycles that began some 36 years apart. The twentieth-century intuitive solution to the problem was mass spraying of the spruce forests. This solution was put into practice and it protected the foliage, but now, instead of cycling every 3.5 decades, the population of spruce budworms threatens to rise well above the economic threshold every year (Watt, 1970), necessitating annual applications of insecticides to save the foliage. Before the use of insecticidal sprays in the spruce forests, the budworm population crashed to very low levels after each eight-year buildup, and did not increase above its economic threshold again for about 36 years; spraying somehow interrupted the natural cycles of budworm populations and kept the population from ever reaching "crash points." Perhaps spraying greatly reduced populations of the natural enemies of the spruce budworms.

Another example: knowledge that many mosaic viruses of plants depend on aphids for their dispersal suggests that insecticidal sprays to control aphids would also prevent the spread of viruses in the fields. But this is not true, and, in fact, insecticidal spraying sometimes results in an increased spread of mosaic viruses. Such viruses are acquired by aphids as they probe superficially in the epidermal cells of diseased plants as a prelude to their feeding in the underlying phloem cells. Viral particles acquired during probes are not ingested; they cling to the stylets, and such viruses are known as the "stylet-borne" viruses. If fields are sprayed with an insecticide, many aphids are killed outright, but some survive the initial blast. Those that survive can acquire and transmit a mosaic virus within seconds. Moreover, spraying disturbs the insects, and the winged forms not killed immediately can fly to nearby plants, which they can then inoculate if their stylets have already been contaminated with virus. Finally, virus-carrying aphids blown into sprayed fields can inoculate protected plants before the protective insecticidal residue on the foliage can kill them.

To reiterate: systems, and particularly complex systems, always possess counterintuitive elements, and these hope-shattering elements must be taken into account by all who would perfect a systems approach to the solution of complex problems. "There is bound to be deception in any approach to the system" (Churchman, 1968). But this Achilles heel of the systems approach neither negates its ultimate effectiveness nor obviates the necessity for it. The systems approach is also characterized by flexibility: the approach can be corrected as tentative solutions to problems are proved wrong. Moreover, the systems approach is characterized by logic and straight thinking, which are indispensable to progress toward the solution of difficult problems. Let us now examine the system approach and see how its principles apply to the problems of plant-pest control.

THE SYSTEMS APPROACH

Hark back to the original meaning of system: to place together. This is the systems approach, which is a striving to put together all the pieces, each in its appointed place, to produce an integrated whole. A system is like a jigsaw puzzle, for each component meshes in only one right way with its adjacent and closely related components. And the approach to the solution of a jigsaw puzzle is sheer logic: first, find the corner pieces; second, interlock the side pieces with the corners and join them together to set the boundaries; finally, complete the puzzle by meshing the remaining pieces. So, with the systems approach, first set the boundaries, and then proceed to solve the problems with all the logic that can be mustered. The following description of the systems approach is based largely on the definitive account of the subject by C. W. Churchman (1968).

Right Thinking at the Outset

Churchman (1968) illustrates how diffficult it is to think correctly at the very beginning, as we undertake for the first time to use the systems approach to solve serious and complex problems. I now paraphrase one of his exercises in thinking: Given the fact that we possess the technology to do so, why do we not solve the problem of world hunger? Because the world is not organized to feed its population; political and economic systems are not yet geared to do so. Why not organize to feed the people of the world? Because any nation strong enough to organize the others would not be trusted. International politics being what they are, efforts to organize could lead to war. Then why not organize international political systems to stop all wars? This cannot be done without educating the peoples of the world. Why not educate the people first? You can't educate a starving man; so the first problem to be solved, it seems, is that of world hunger. And we're right back where we started, having accomplished nothing on our travels along this circuitous pathway of the mind. We asked the wrong question to begin with.

How does one get on the right track at the beginning? There is no sure way, but at least two principles of the systems approach point in the right direction. First, the systems approach is a way of thinking, and one must start with specific thoughts about small subsystems, and precisely state a specific objective. Clearly, the principle of specificity was ignored in the exercise we just looked at. Second, the systems approach is not so much one of form as one of function. This principle was also ignored in the exercise. To quote Churchman (1968): "We may have begun incorrectly because we began by describing the world in terms of its structure, not its purpose." Right thinking at the outset is asking the question: What, specifically, are we trying to do?

The Structure of Systems

Right thinking about systems at the outset is thinking in terms of function, but as a mental image of a functioning system develops, that system takes on form, form that is structured to accommodate function, to make the system work. Five elements give any system its form. A system consists of objectives, environment, resources, components (missions), and management. And as a system is managed, a somewhat different and more sophisticated view of its structure emerges. Thus Forrester (1969) enumerates four hierarchies of systems structure: a closed boundary, a network of feedback loops, level (state) variables accumulating within feedback loops, and rate (flow) variables as measures of the speed with which activities proceed in the feedback loops. Finally, successful management of a system so characterized depends on cybernetics (steersmanship), which has its own peculiarities and structure (Ashby, 1956). These aspects of the structure of systems are treated in this section, in which I review briefly the concepts already detailed by the authors previously cited in this chapter.

Objectives and Boundaries

The structure of any system endures or decays on the strength or weakness of its objectives. The successful planner of a system starts right: he thinks specific thoughts about what the system is supposed to do, how it is to work. He sets specific boundaries, within which the system can function, recognizing that his system will interact with other systems, which in turn will also affect his own modest system.

Consider a relatively simple system: a system for undergraduate education in crop-plant-pest control. The overall objective, it seems, would be to educate and train college-level students in the science and some of the arts of plant-pest control. This objective is one of function: it states the purpose, not the form, of the system. Also, it is specific, though it may not seem so at first glance. But the boundaries are clearly drawn. The educational program envisioned is for baccalaureate students only. It goes beyond the simple training of observers or "scouts," which training would require no more than a grade-school education, a desire to learn, and a willingness to work. It stops short of the specialized education of graduate schools, whose job is to educate (and train) the professionals of the several disciplines that treat the different aspects of plant-pest control. The stated objective also sets another kind of boundary: it limits the system to the treatment of crop-plant pests, thus excluding household pests, pests of human beings, and pests of animals, however important those other pests may be.

The specificity of the objective of this hypothetical system of education becomes apparent as we examine several goals to be sought in the pursuit of the overall objective. What are some of these goals? We seek to make students aware of plant-pest control as a science; we encourage them to master the principles and concepts of the science; we have them learn to think in terms of integrated systems of plant-pest control; we give them limited apprenticeships so they can practice a few acquired skills; and so on. But all specific goals must somehow help attain the overall objective, and any specific goal perceived to fail in this endeavor would be knowingly, willingly, and speedily sacrificed.

The Environment of the System

Environment, in the context of the systems approach and systems structure, is that which lies outside the system; the system cannot do very much about its environment, but the environment often determines how the system works. One must distinguish between the environment outside a system and the environment encompassing the components within a system. For the hypothetical system of undergraduate education in crop-plant-pest control, the environment is composed of several hierarchies of administration, none of which is much affected by the system. These hierarchies include administrative units of the college, the university, the trustees or regents, and, in public colleges and universities, the state legislature. The environment within the system (class study materials, teaching aids, and the like) is very different: it is an integral part of the system itself and is subject to revision as the system is managed in order to accomplish its missions and thereby achieve its overall objective.

Having stated objectives in terms of function, and having taken note of what lies outside the system, we turn now to the heart of a system at work. Consider, next, the resources and the components of systems.

Resources

"The means that the system uses to do its job are its resources" (Churchman, 1968). Resources include personnel, money, and equipment, and are, therefore, the tools of the system. All the resources interact as they are used to complete the missions that contribute to the attainment of the objective. And, perhaps most importantly, the system can change its resources as they are perceived to be lacking in any significant respect.

To continue with our example: What are some of the resources within the system of undergraduate education in crop-plant-pest control? Personnel: professors thoroughly educated and highly skilled in each of the several disciplines that deal with plants, at least one professor of epidemiology, and one familiar with the details of systems analysis and the systems approach. Adjunct professors of plant physiology and ecology, of agronomy, of soil science, of horticulture, and of forestry, it seems, would also serve as resource personnel. Then there is the vital resource, money: money to pay the staff, money to provide a suitable environment within the system (classrooms, laboratories, field plots, etc.), money to purchase other important resources, such as study materials and teaching aids. And, in a system for education, the curriculum is a central resource. Within our hypothetical system might be such courses as elementary plant-pest control, specific courses on the nature and control of each of the several kinds of plant pests, an internship, a seminar, and, to bring to all together, a course on the systems approach to plant-pest control. Every resource of the mission is then used as an instrument to perform the specific task for which it was designed. Such tasks, or missions, are the components of the system.

Components

The functional units of a system constitute its components, and by functional units one means activities, specific goals—in a word: subsystems. In our system of college education in crop-plant-pest management, for example, the student is the central component, but other components might include such units of study as world crops and losses due to plant pests, the form and function of crop plants, agricultural ecology, population dynamics, or principles of plant-pest control. Moreover, certain resource persons would serve also as working components of the system. The classroom teachers, for example, play a dual role: they are certainly a teaching resource, but they become also a component of the system as they work toward a specific goal and observe the results of their activities within the system. Components and resources that work reassure management of the soundness of the system's objective; those that do not work (those that are "counterintuitive") convince management that a change is necessary. Management is the fifth element of any system.

Management

Management is responsible for the system in all matters relating to its plan. Management first makes the plan, formulating the overall objective and even stating all possible specific goals within the main objective. Management must provide information and evaluation systems to find out if the plans work. Finally, management must provide machinery that would facilitate a change in plan if the information gathered from within the functioning system indicated the need for change.

All this interchange of information between the management, resources, and components is accomplished by an interrelated series of information cycles, "feedback loops," which are the basic structural elements within the closed boundary that surrounds every working system (Forrester, 1969). Feedback is a term originally applied to self-regulating machines such as the thermostat. If the thermostat is set at 68°F and the temperature drops below that setting, the information flows to the switch on the furnace and that switch turns on the heat. Later, when the temperature reaches 68°, information to that effect flows to the switch, which then automatically changes to the off position, and the furnace is turned off.

The successful management of a system depends on recognizing the two kinds of variables within each feedback loop and on comprehending the significance of whatever information accumulates and of how fast it flows from one variable to another. What are thse two variables, which are the third and fourth hierarchies of Forrester's (1969) view of system structure? They are the state variables and the rate variables that operate within every feedback loop. "Information" accumulates to a certain point or level, and a state or level variable results. The state (level) tells how much of something there is at a given time: what the temperature is, how much money is in the bank, how much of a pest population is in the field. The rate (flow) tells how fast the state variable changes and in what direction it changes: whether the temperature changes and how much it rises or falls, how much money was deposited or withdrawn over a period of time, how much the birth rates and death rates of plant pests changed with time. And the state variable changes only as the rate of information flow into it or away from it changes. Management seeks to control the flow rate and does so after evaluation of the state or level variable. For example, if the bank balance is low, withdrawals are curtailed and (if possible) deposits are made. If the population of a plant pest threatens to exceed the economic threshold, a control measure may be put into practice to decelerate the rate of build-up of the pest population.

The simplest feedback loop operating within the boundaries of a system (Figure 11.2) consists of a straight-line flow of "information" in the system to a state or level variable, in which the information accumulates. Along the way, though, the information is subjected to acceleration or deceleration by the action of the rate or flow variable. Finally, information is cycled back to the rate variable from the state variable, and, on the basis of this information, a decision is made that controls the flow or action stream (Forrester, 1969).

Feedback loops and their two kinds of variables can be illustrated by one facet of our recurring system of education. One state variable, for example, would concern the accumulation of knowledge and the mastery of skills by the students, who are quite variable in their receptiveness and attentiveness to the flow of information to them, and very different in their abilities to retain information and to perceive the practical applications thereof. But a certain level of achievement becomes apparent after a while, particularly after an examination. Then the professor, as manager of the system, corrects the original plan, by manipulating the rate (flow) variables, which involve such matters as methods of presenting information (lecture, reading assignments,

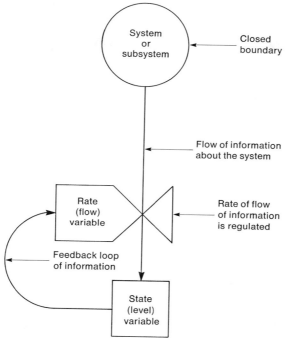

FIGURE   11.2
A simple feedback loop.

demonstrations, group discussions, etc.) and the rate at which the several units of information are presented.

Management thus develops plans, identifies functional feedback loops of information, provides for an evaluation of the performance of the system, and promptly changes its plans if its information so indicates. The manager of a system holds in his hands a dynamic thing, ever changing, constantly in motion. The manager has been likened to the steersman of an ocean-going vessel, who must constantly take action and counteraction if he would steer a sure course to reach port safely and according to plan. Management of a system is thus "steersmanship," and that fine art rests upon the principles of a new science termed cybernetics.

Cybernetics

The science of cybernetics deals with the mechanisms of control and transmission of information. Weiner (1948) coined the term cybernetics, and proposed that control and communication in engineering and in biology be studied together. Cybernetics is "the art of steersmanship," and has two peculiarities: first, it treats ways of behaving rather than things; and, second, it permits the scientific treatment of systems whose complexity is awesome but too important to be overlooked (Ashby, 1956).

The idea of the systems approach is to deal with complexities, and systems management seeks to enumerate and then come to understand all variables within the system. But, obviously, not every variable becomes apparent; there will always be some gaps in our knowledge of a complex system. This realization need not be a deterrent against the systems approach. First, as the system is analyzed, new facts and relationships will be perceived. Second, unknown variables within a system can be treated together as the unseen contents of a "black box." Without understanding its contents, managers can learn by trial and error how the mysterious variables within the "black box" interact with the known components of the system.

After two centuries of exploring the simplest of systems, where a single factor is allowed to vary as all others are kept constant, science is finally beginning to study complexity itself as it employs the methods of cybernetics (Ashby, 1956). And within the structure of systems, there is a sequence of steps to be followed as the systems approach is applied to complex problems. These steps characterize the chronology of systems research and constitute its technology.

Chronology and Technology of the Systems Approach

We come now to the end-result of the systems approach: the putting together of facts and ideas into a practical scheme for achieving the goals of the system.The application of systems thinking to everyday problems is exemplified by systems analysis.

Systems Analysis

The technology of the systems approach is systems analysis, whose chronology is punctuated by four sequential events: measurement, analysis, description, and simulation (Watt, 1966; Zadoks, 1971).

First, all perceived variables—rate variables as well as state variables—must be measured. For example, the analysis of an epidemic of even a single plant disease would demand measurement, at intervals of time, of such rate variables as rainfall, temperature, duration of wetting, and other factors that might affect such a state level as the population of the disease-causing pest. And the level of the pest population certainly must be measured—often indirectly—at numerous intervals during the development of the epidemic. As with all science, systems analysis begins with keen observations and meticulous recordkeeping.

Second, during analysis proper, quantitative relations between the measured variables are determined. The analysis of an epidemic, for example, seeks to relate the percentage of crop loss during a period of time to the fluctuations of rate variables during the same period of time. Though favorable to disease development, short-term meteorological conditions do not necessarily make for an epidemic. Favorable factors must persist for a specified period of time if they are to permit a widespread outbreak of disease.

Third, the systems analyst sets down a description of the system as he is able to construct it from earlier measurements and analyses. Description, it seems, is an expository account of the system: detailed, but not necessarily critical or highly developed. Description is the basis for the last step in the analysis.

Finally, the systems analysis is completed by simulation. In this step the already-described system is fully developed into a working model of the real system. Simulation is the construction of a model and also involves the study of its behavior (Zadoks, 1971). The measurements and analyses are finally put together, usually with the aid of a computer programmed with the data collected earlier in the analysis. Once the computer model has been constructed, the systems manager can finally put his concepts and perceptions of the system to the test. The model should perform like the real system, and if it does not, then errors in measurement, analysis, description, or simulation must exist. Once the errors have been identified and the flaws corrected, the model can be reconstructed according to "new" information. It will then more nearly approach mimicry of the real thing. Working models thus permit systems managers to correct their early mistakes, to change their faulty plans for systems, and to adjust their approaches to complex problems accordingly.

The management component of a system can function only if the system is sufficiently flexible to permit abrupt changes in plan if the feedback loops yield information indicating the necessity for a change. How the technology of the systems approach might apply to the many facets of plant-pest control is strikingly illustrated by analyses of plant-disease epidemics.

An Application of the Systems Approach
in Plant-Pest Control

Although the systems approach is holistic, it is so—at least for the individual scientist—only for the smallest of subsystems within the larger ones. One who would make a significant contribution to our knowledge of integrated plant-pest control, then, must make the right beginning: he must start with straight thinking about a subsystem whose intricacies he has already mastered or has a reasonable chance of mastering as he proceeds with his analysis of the system.

Waggoner and his associates at the Connecticut Agricultural Experiment Station are among the pioneers in computer simulation of plant-disease epidemics. They have developed working models for the late-blight (Waggoner, 1968) and early-blight (Waggoner and Horsfall, 1969) diseases of potatoes and for the southern leaf blight of corn (Waggoner et al., 1972). These models serve as examples of the kind of work much needed in all areas of plant-pest control. Let is now examine one of these models, EPIDEM, which simulates the early-blight disease of potato, caused by the fungal pest *Alternaria solani.*

First, Waggoner and Horsfall (1969) practiced right thinking from the outset: they elected to study a relatively small complex system, a fungal disease of a crop plant. It was a system they could manage.

Second, they practiced the systems approach. Their overall objective was to make the selected disease comprehensible, so that threatening epidemics could be predicted and the familiar control measure of protective spraying could be practiced when needed, but only when needed. They were cognizant of the environment outside the system, but did not waste time on extraneous matters little affected by their system. Their resources included personnel, published information, facilities to perform needed experiments, and the tool (computer) for simulating the epidemic. They studied the details of the two principal components of their system, the life cycle of the fungal pest and the environmental factors within the system that affected each event in the life cycle of the fungus. Finally, they managed their system, practicing cybernetics as changes in plan were indicated, and culminated their management by a thorough analysis of the system. In that analysis, they measured variables, analyzed their interactions, described their system with words, and finally translated their words into computer language and developed a simulation model of the real epidemic. EPIDEM thus emerges as a splendid example of the classical systems approach to a complex problem of plant-pest control.

The technological details of computer simulation—the use of Fortran or another computer language, higher mathematics, and the like—are beyond the scope of this book, whose purpose is merely to introduce the principles of the systems approach. Nevertheless, some of the detailed information derived from the development of EPIDEM (Waggoner and Horsfall, 1969) seems worthy of mention

TABLE   11.1.

Relations between weather and events in the life cycle of *Alternaria solani,* the agent of early blight of potato.[a]

Event	Favorable weather
Spore-stalk formation	Light, warm, wet (but no rain)
Spore production	Dark, cool, wet
Spore dispersal	Dry, heavy rain, brisk winds
Inoculation (spore landing)	Light wind, light rain
Spore germination	Humid air, wet surface
Penetration	Cloudy, warm, wet
Infection (colonization of leaves)	Cool, wet

[a]*Source:* Waggoner and Horsfall (1969).

here, for such information dispelled much of the confusion that had enshrouded the early-blight disease of potato. Depending on their sources of information prior to EPIDEM, scientists regarded the disease as one of warm weather or of cool weather, one of wet weather or of dry weather. On the basis of their analysis, Waggoner and Horsfall (1969) concluded: "*Alternaria,* like people, likes a variety of weather in the right season." This is clearly indicated by a listing of meteorological factors favorable to each of several events in the life cycle of *Alternaria solani* (Table 11.1). Warm weather favors the development of spore stalks, but cool weather favors spore formation; most events are favored by wet weather, but spore dissemination— essential to the epidemic—is favored by dry weather. Little wonder, then, that early blight of potato was a source of considerable confusion (and argument) before everything was brought together in EPIDEM (Waggoner and Horsfall, 1969).

   The success of EPIDEM and other programs such as EPISIM (Zadoks and Rijskijk, 1972) and EPIDEMIC (Shrum, 1975) provides reason enough, it seems, to employ the systems approach to solve the complex problems of plant-pest control. Despite its complexity and its counterintuitive nature, "the systems approach is not a bad idea" (Churchman, 1968).

LITERATURE CITED

Ashby, W.R. 1956. *An Introduction to Cybernetics.* London: Chapman and Hall.

Churchman, C. W. 1968. *The Systems Approach.* New York: Delacorte.

Forrester, J. W. 1969. *Urban Dynamics*. Cambridge, MA: M.I.T. Press.

Shrum, R. 1975. "Simulation of wheat stripe rust *(Puccinia striiformis* West.) using EPIDEMIC, a flexible plant-disease simulator." Progress Report 347, Pennsylvania State Univ., University Park.

Waggoner, P. E. 1968. "Weather and the rise and fall of fungi." In W. P. Lowry, ed., *Biometeorology*. Corvallis: Oregon State Univ. Press.

_____, and J. G. Horsfall. 1969. "EPIDEM, a simulator of plant disease written for a computer." *Connecticut Agr. Expt. Sta. Bull.*, no. 698.

_____, J. G. Horsfall, and R. J. Lukens. 1972. "EPIMAY, a simulator of cornleaf blight." *Connecticut Agr. Expt. Sta. Bull.*, no. 729.

Watt, K. E. F. 1966. *Systems Analysis in Ecology*. New York: Academic Press.

_____. 1970. "The systems point of view in pest management." In R.L. Rabb and F. E. Guthrie, eds., *Concepts of Pest Management*. Raleigh: North Carolina State Univ. Press.

Weiner, N. 1948. *Cybernetics*. New York: Wiley.

Zadoks, J. C. 1971. "Systems analysis and dynamics of epidemics." *Phytopathology*, 61:600-610.

_____, and F. H. Rijskijk. 1972. "Epidemiology and forecasting of cereal rust: Studies by means of a computer simulator named EPISIM 1972." *Eur. Med. Cereal Rust Conf.*, Prague, pp. 193-196.

12

Epilogue: Plant-Pest Control in World Agriculture

Plant pests, as they proliferate their populations at the expense of the green plants that feed, clothe, house, and inspire mankind, behave as retarding factors; they decelerate the rates at which crop plants manufacture their useful products. And because every growing season unfolds, develops, and terminates within unyielding boundaries of time and space, rate-retarding factors are finally made manifest in terms of reduced crop-plant yields. Control measures retard the rates of population increase among plant pests, but plant-pest control, in the general sense, can be viewed as a limiting factor for crop-plant production. That is, as pest control is practiced, rates of plant production accelerate, thus accounting for increased yields before the growing season ends.

But plant-pest control, as a limit to some degree on crop-plant production, represents only a tiny element in the sum of modern agriculture. Plant-pest control ranks low in a hierarchy of agricultural systems that people manage within complex natural ecosystems. Above it are general systems of agriculture that encompass other subsystems along with that of plant-pest control. Among the other subsystems are those of plant production, plant use, and the distribution of plant products. Nevertheless, none can doubt that plant-pest control is essential in modern agriculture. Without it, world crop production would plummet to abysmal depths, for any one of a myriad of plant pests could, if unchecked, totally destroy a crop. It is the potential damage threatened by plant pests that we must somehow avert.

Although it ranks as only a subsystem within agricultural systems, plant-pest control is complex in the extreme. And it embraces a series of additional systems, each about as complex as the other. There are thus complex subsystems for each major group of plant pests: vermin, weeds, harmful insects, and plant pathogens. Then there are sub-groupings within each of the major groups. Finally, every pest-victim interaction makes up a complex system within the hierarchy. Little wonder, then, that the subject of plant-pest control doggedly defies human comprehension. Some understanding of plant-pest control, however, can be achieved by the systems approach, which justifies a cautiously optimistic view of the future. Let us now try to put the sciences and the arts of plant-pest control in agricultural perspective. Let us take the systems approach: start with right thinking by deciding first what we are trying to do; think specific thoughts about subsys-tems whose intricacies we can master; and finally, for every system, let us clearly state its objective, learn about its environment, marshal its resources, work with its components, and manage the system by the practice of cybernetics. We must exercise patience, expect the unexpected, and rectify early mistakes. In this chapter, I treat systems of plant-pest control as they might be practiced in the industrial and agrarian countries of the world.

PLANT-PEST CONTROL
IN INDUSTRIALIZED COUNTRIES

The impact of the Industrial Revolution has been felt as much by the farmer as by any other segment of society in the industrialized coun-tries. And the rate of change in farm-management practices wrought by the Industrial Revolution has accelerated exponentially, it seems, during the past three or four decades. Whereas the successful farmer in the United States of the 1920's could practice his arts and skills very much as his father and grandfather had before him, the present-day farmer must use production procedures developed as recently as a few years earlier, in order to compete with progressive neighbors. Of all the new farm practices, perhaps none is as new as that of integrated pest control. Intense pressure is thus brought to bear on agencies that educate—in the colleges and on the farms—for plant-pest control, and on agencies that conduct the experimental research that leads to rec-ommendations for effective plant-pest control on an integrated basis.

Education and Training for Plant-Pest Control

Those engaged in plant-pest control, if they would keep abreast of important developments in an every-changing agricultural world, must learn continuously, from an early age and throughout their careers. And the institutions of our society owe it to the practitioners of plant-pest control to offer opportunities for learning at every level of understanding. In an industrialized country such as the United States, for example, secondary schools and colleges might teach fundamentals of plant-pest control and the Cooperative Extension Service might undertake the task of providing opportunities for continuing education on the farms, where plant-pest control is put into practice and where the principles mastered in school are put to the test.

Education in Schools and Colleges

Perhaps we should, at the outset, attempt to distinguish between education and training. The two concepts share common characteristics: both involve teaching and learning; both demand strict discipline; both are essentially formative in nature. But education develops primarily the mental skills, whereas training develops the physical or manual skills. For the subject of plant-pest control, we can view education as the teaching and learning of its sciences, training as the teaching and learning of its arts. And the expert in plant-pest control is well educated and highly trained; educated mostly in the schools and colleges, trained mostly on the job.

What knowledge might the well-educated plant-pest-control specialist acquire from such schooling? First, of course, he (or she) develops an acute awareness of plant-pest control as an integrated multidisciplinary science. He learns the major taxa that include plant pests. He learns something of the nature of each very different kind of plant pest; he knows the enemies of the crop plants he would grow. And he learns about his crop plants, what they look like and how they work. He learns how crop plants function as members of agricultural systems within natural ecosystems. He masters the principles of plant-pest control; that is, he learns the laws that govern pest-victim-environment interactions. And he approaches his complex plant-pest-control problems from the systems point of view.

Such knowledge should be available to would-be plant-pest-control experts in our schools and colleges. To what lengths the student might go to attain an education in the subject depends on the time, effort, and money he or she wishes to invest in quest of an education in plant-pest control. For example, the contents of this book offer a first-level understanding of the subject, suitable for beginning college students. College graduates, it seems, should also study, in some detail, plant pathology, plant-disease control, entomology, insect control, weed science, weed control, vermin and their control, and systems of plant-pest control. Moreover, training in the skills of plant-pest control might well begin with a summer job under the supervision and tutelage of a practicing plant-pest-control specialist. Finally, the student could terminate formal education at the graduate level, specializing in one of the smaller subsystems that he could expect to manage and whose subject matter he could expect to master. Is such education offered in the schools and colleges of the United States and other industrialized countries? The answer is a qualified yes. Some already have curricula that provide education for plant-pest control, and many others will soon join their sister institutions. But not all who must practice plant-pest control can avail themselves of opportunities to become educated in schools and colleges. What of the practicing farmer? Though trained through experience and highly skilled from much practice, he or she would still benefit from education and training in up-to-date approaches, such as is offered by the Cooperative Extension Service.

Education and Training on the Farm

Of necessity, perhaps, farm advisory services have devoted much time to trouble shooting, and have frequently considered their duties done after identifying a specific problem and relaying a specific remedy recommended by a specialist at the national or state (provincial) level. Trouble shooting is important, and will always be important in plant-pest control, as it is universally. When the tractor breaks down, the trouble must be located and the machine must be fixed. But farm machinery works best when given regular maintenance. Is it too much to demand the same for plant-pest-control procedures?

The successful farmer plans ahead. For plant-pest control, he must make hard decisions about preplanting treatments of his soil, his choice of crop-plant varieties, seed treatments, control measures likely to be called for during the growing season, and procedures required after the harvest. In short, he practices the systems approach. But most present-day approaches to plant-pest control, even in the industrialized countries, are found wanting. Few plans take into account all the possible pest problems already known to science. Even fewer, it seems, consider the recognized interactions among the subsystems within the larger systems. The overall objectives of the advisers in plant-pest control, then, might be to demonstrate to the growers of plants the scientific bases for effective pest control and to assist farmers in development of their own systems for integrated plant-pest control. But the scientific bases for integrated plant-pest control rest on the discoveries of researchers, whose efforts bring forth the knowledge that teachers can impart and that advisory people can relay to the practitioners.

Research on Plant-Pest Control

All scientific research proceeds within the framework of the familiar scientific method: state the problem; develop hypotheses intended to solve the problem; test the hypotheses with critical experiments and observe the results; interpret the data; and draw conclusions made valid by those data. In terms of function, however, research is of two very different kinds: fundamental (basic or pure) and applied. The distinction between the two is not always apparent, but fundamental research seeks knowledge; applied research seeks to discover the usefulness of that knowledge in everyday life. For example, a study of the genetics of disease-resistance or insect-resistance in pure lines of a crop-plant species is fundamental research; a project to breed and select a new variety for resistance is applied research. Both kinds of research are indispensable to plant-pest control: applied to solve the problems of the moment, fundamental to give us the bases for solving the problems of the future.

The literature on plant-pest control is already voluminous, and innumerable volumes are yet to written. But there is a paucity of literature on experiments that have sought to put the science all together. We have experimented much with individual plant pests and their victims, and some with populations of specific pests. But we have experimented relatively little with the interactions among populations of all the different kinds of plant pests and their diverse victim species. It is time that we expend every effort to put it all together. Who can do it? Those already established as competent researchers in wildlife management, weed science, entomology, and plant pathology can devote their energies to the task. But first they must begin by thinking beyond their own academic disciplines; they must think on systems of integrated plant-pest control. Then they must collaborate in their researches, because no individual specialist can be expected to comprehend the mysteries of all the sciences that relate to the subject.

Coordination of Systems of Plant-Pest Control

Research, advisory work, and classroom instruction in plant-pest control must begin and end with the individuals engaged in those activities. They must come together before they can put it all together. They must do so voluntarily, which means that they must first be of a mind to do so. But another ingredient, management, must be blended in with the other components of the whole system. Although some of the research, extension, and teaching personnel must shoulder the burden of leadership, administrative personnel of experimental units (private as well as governmental), extension services, and schools, colleges, and universities should create an environment favorable to the programs of work and study upon which a successful program of integrated plant-pest control depends. Coordinated programs have sprung up throughout the industrialized regions of the world during the past two decades, and the burgeoning interest in this important facet of agriculture augurs well for the future of plant-pest control in the industrialized and agrarian countries.

Careers in Plant-Pest Control

The educated and trained specialists will manage the coordinated programs of integrated plant-pest control alluded to in the preceding section. It is they who will make the day-to-day decisions that will keep the system working. It seems likely that specialists educated in four-year colleges will manage most plant-pest-control systems in the field, and it seems appropriate, therefore, to consider the kinds of jobs the college graduate can expect to find.

Specialists in plant-pest control are much in demand to fill a variety of positions in educational organizations, in commercial firms, on the farms, and in governmental agencies. They may teach in schools and community colleges, or they may assist professors in college courses. They may work as technologists and assist in the conduct of research by professionals in universities, agricultural experiment stations, and in the United States Department of Agriculture. Graduate specialists may practice their technology in the laboratory, greenhouse, and field, and some may work at computer terminals as more and more systems of pest control are subjected to systems analyses.

Teaching and research jobs also await those who prefer to work in extension services. County and district agricultural agents advise farmers, instruct them, and conduct applied research on the farms with the cooperation of the farm managers. Another kind of career, that of information specialist, might be attractive to the graduate with a flair for writing. The field of agricultural journalism is open to good writers who are also versed in technical subjects.

Commercial firms dealing in agricultural goods, products, and services offer many career opportunities to graduate specialists in plant-pest control. These organizations seek reliable technologists to conduct research, knowledgeable graduates to work in technical-sales departments, and energetic specialists to serve as salesmen and advisers to farmers. Moreover, companies that contract to control pests on farms and estates, in cities and parks, and along rights-of-way are in the market for graduate specialists.

The individuals and corporations engaged in growing crop plants for products and profit also hire specialists in plant-pest control. They seek competent observers and skilled pest-control practitioners in their efforts to avert pest-inflicted damage in their fields, orchards, or forests.

The practices of plant-pest control proceed within a framework of regulations intended to protect agriculture and the consumers of its bounty. National and state agencies that enforce regulations rely heavily on the observations and recommendations of the graduate pest-control specialists they employ. Regulatory agencies hire specialists as inspectors to work in the fields, orchards, and nurseries and at points of entry into geographical and political subdivisions. They hire specialists who must make hard decisions about embargoes and quarantines and who can convincingly instruct legislative and judicial bodies in matters of plant-pest control.

These, then, are the kinds of jobs the graduate specialist in plant-pest control might undertake, whether working for an employer or for himself. In any event, he or she would in some way be engaged in controlling the pests of plants, and in so doing, would be subject to certain constraints imposed by law.

Some Legal Aspects of Plant-Pest Control

Laws governing the distribution of plants and plant products guard against the introduction of undesirable plants and the inadvertent introduction of pests that might infest or infect them. These laws, enforced by regulatory agencies, are well-understood and seem to be generally accepted by scientist and layman alike. Adherence to quarantine laws is sometimes costly and often inconvenient, but these laws seem not to have sparked such controversy as have the recent laws governing the use of pesticides. I comment in this section on the Federal Environmental Pesticide Control Act of 1972 and its predecessors.

Five landmark acts of legislation and five amendments to those acts punctuate the twentieth-century chronology of laws passed in the United States, laws enacted primarily for the protection of the consumer and for the preservation of the environment. First, the Pure

Food Law of 1906 demanded that the food products of interstate commerce be pure and wholesome, but it disregarded pesticidal residues on or in foodstuffs. The amendments of 1938 and 1954, however, dealt specifically with pesticides that affect food: the first amendment established levels of tolerance for pesticides; the second required clearance (by the Food and Drug Administration) attesting to the safety of products contaminated by residual amounts of pesticides. A third amendment in 1958 applied to pesticides not covered earlier and permitted no residue of chemicals that are carcinogenic in laboratory animals.

Second, the Federal Insecticide Act of 1910 (which dealt also with fungicides) was enacted to protect the users of certain pesticides against fraudulent claims of the manufacturers and distributors. This law was the forerunner of a third piece of pesticidal legislation, the Federal Insecticide, Fungicide, and Rodenticide Act of 1947, which applied also to herbicides. It required registration and complete labeling of pesticides. This law was amended twice: in 1959 to include nematicides, plant-growth regulators, defoliants, and desiccants; and in 1964 to have the federal registration number on all labels and to have words of caution conspicuously printed on labels of containers of pesticides poisonous to humans. The Pesticides Regulation Division of the United States Department of Agriculture administered the 1947 law until 1970, when that responsibility was assigned to the Environmental Protection Agency, which came into being upon passage of a fourth piece of legislation, the National Environmental Policy Act of 1969.

Pesticide legislation in the United States has been recently revised and substantially amended by a fifth law, the Federal Environmental Pesticide Control Act of 1972. This significant legislation states, among other things, that a registered pesticide must be used only according to instructions on the label; every pesticide will be classified for "general" or "restricted" use; pesticides in the "restricted" category may be applied only by applicators who have previously been certified (licensed) by state regulatory agents; all pesticides must be registered; and the manufacturer must submit experimental data attesting to (1) the effectiveness of the pesticide for its stated purpose, (2) its safety to plants, humans, animals, and the environment, and (3) the fact that no harmful residues will remain in foods for humans or in feeds for animals.

The preceding account and the comments on pesticide legislation in Part II of this book refer only to laws of the United States. Similar laws regulating the manufacture, distribution, and use of pesticides are in force elsewhere, particularly in the industrialized countries, but in many agrarian ones as well. And pesticide legislation, I believe, will soon become an integral part of the codes of International Law. A step in this direction was taken in 1972 at the United Nations Conference on the Human Environment. The conferees called for international programs of pest control and recommended ways to inaugurate such programs.

PLANT-PEST CONTROL IN AGRARIAN COUNTRIES

To classify the countries of the world in only two categories, industrialized and agrarian, is certainly an oversimplification and is probably an error. There are primitive features of industrialized countries and highly civilized features of agrarian countries. Still, there are clear differences in general between industrialized and agrarian countries, by which I mean developed and undeveloped (developing, emerging) countries, respectively. The Industrial Revolution has not yet had its full impact on the so-called agrarian countries: machines have not yet replaced manual labor. The roots of the problem of world hunger are deeply imbedded in these countries, and it is there that systems of effective plant-pest control are likely to have their most far-reaching effects.

But whether a country has become highly industrialized or remains mostly agricultural, the objective of plant-pest control is the same. What we are trying to do is to keep the population of plant pests at or below their economic thresholds. The world is thus united on at least one point, significantly, a point of function, not one of form. This unity of purpose means that the principles of education, training, and research mentioned in the first section of this chapter apply to systems of plant-pest control in the industrialized and agrarian countries alike. Differences are expressed in form, the ways in which we tackle the problems and struggle toward the goal. Similarly, the principles of plant-pest control defined in Chapter 5 are universal; only the application of control measures exemplifying those seven principles varies from place to place and from time to time.

How might systems for plant-pest control evolve in agrarian countries? First, the successful system develops from within, within the district, the province, the state. It evolves by means of the efforts of personnel who have been trained to identify the problems in order of their priorities and educated in the principles of plant-pest control and in the systems approach. It evolves also by means of the efforts of academic and political leaders who have been educated to an awareness of plant-pest control and its significance for agriculture. And where do the personnel receive their training and obtain their education? In their own countries, insofar as that is possible, but in the industrialized countries if necessary. Developed nations have an obligation to undeveloped nations, but that obligation, I believe, is one of helping others to help themselves. Perhaps industrialized countries can assist in the education and training of plant-pest-control specialists who will serve in their own agrarian countries. Perhaps advisers on leave from industrialized countries can assist leaders of agrarian countries in the planning of systems for plant-pest control. Perhaps workers in developed countries can provide germ plasm of potentially desirable crop-plant varieties for testing in undeveloped countries. But in the final analysis, the age-old wisdom of Aesop is fulfilled: "The gods help them that help themselves."

PERORATION

The remarks made in this epilogue are in no way intended to suggest a universal structure within which systems of plant-pest control might be practiced. They are intended, instead, to urge a unity of thought in terms of function. Having a knowledge of crop plants and recognizing the noxious organisms that beset them, knowing the seven principles that undergird the diverse measures of plant-pest control, comprehending the systems approach to complex problems, and possessing the determination to "put it all together," we find ourselves in the position of forging an unbreakable chain of knowledge of plant-pest control, a chain that links scientist, teacher, adviser, administrator, and farmer to the Earth, and ensures that Earth remains forever the planet of green plants.

Index

Abiotic agents of plant diseases, 92
Abutyramide, 199
Acarine families (mites), 148
Actinomycetes, as infectious agents of
    plant diseases, 95
Adaptation, 192
Aerobic respiration, 42
Agrarian countries
    characterized, 231
    systems of plant-pest control in,
        231–232
Agricultural systems, 223
Agriculture 3, 4
    systems of, 223
Agrobacterium, characteristics of, 25
Alarm sounds, to frighten birds, 194
Alfalfa, classification of, 23
Alfalfa weevil, 153
Alternaria solani, 220, 221
Alternation of generations, 37
Amensalism, 66
Anabiosis, in nematodes, 130
Anabolic reactions, defined, 41
Anatomy of plants
    as a basis for the classification of
        plants, 24
    defined, 21
    as being suggestive of plant functions,
        24
Angiospermae, characterized as a class
    of plants, 26
Anguina, as a sedentary endoparasitic
    nematode, 123
Animals, domesticated, 4
Anoplura, characterized as an order of
    insects, 141
Anther, 34
Antibiotics, resistance of plant-parasitic
    bacteria to, 76
Aphelenchoides besseyi, 123, 137
Aphids
    control of by sanitation, 157
    feeding on a plant, 147
    as homopterous insects, 141
    repelled by aluminum mulch, 157
    as transmitters of mosaic viruses, 210
    vertical resistance of corn to, 77
Appendages of the plant body, 31
Apple, classification of the plant, 23

Applied entomology, defined, 139
Applied methods of insect control,
    156–161
Applied research, distinguished from
    fundamental research, 227
Armadillos, controlled by trapping, 202
Arthropods, as plant pests, 11, 142
Asparagus, toxic to nematodes, 133
Atrazine, as an herbicide, 178
Autecology, defined, 13
Auxins
    in the control of plant growth, 53
    as growth regulators in roots, 50
Avitrol, as a repellent of birds, 196, 198
Avoidance
    of depredations by birds, 199–200
    environment and, 81
    in insect control, 161
    in nematode control, 137
    as a principle of plant-disease control,
        103, 112
    as a principle of plant-pest control, 81
Awl nematode, as an ectoparasite, 123
Axis of the plant body, parts of, 27, 28

Bacillus, characteristics of, 25
Bacteria
    nitrifying, harmed by nematicides,
        133
    number of plant-parasitic species, 95
Bald Eagle Act, 195
Banana
    classification of, 23
    nematode control in corms of, 129
Barium carbonate, for vermin control,
    189
Bark beetles, 157
Barrier crops
    in the control of plant viruses, 113
    experiments with, 192
Bats (genus *Pteropus*), banned from the
    United States, 200
Bean
    classification of, 23
    leaves damaged by the Mexican bean
        beetle, 144
    root-knot resistance in varieties of, 129
    seedling of, 49
    seed of, 36

233

Beavers, controlled by trapping, 202
Beetles
    as coleopterous insects, 141
    Mexican bean, 144
Belonolaimus
    as an ectoparasitic nematode, 123
    longicaudatus
        controlled by nematicides, 132
        effect on citrus, 136
        strains of, 134
Beggarweed, 165
Benomyl, as a protective and
    therapeutic fungicide, 116, 118
Bensulide, as an herbicide, 178
Beta vulgaris, classification of, 23
Biological control
    of insects, 155
    of plant parasites, 110–111
    of vermin, 190
    of weeds, 174–176
Biomass, defined, 13
Biotic potential, 62, 63
Birds, as plant pests, 194–195
Birth rates, 61
"Black box," 218
Blackbirds, behavior of, 196
Blackspot disease of rose, 92
Blast, as a hyperplastic symptom of plant
    disease, 91
Blight, as a necrotic symptom of plant
    disease, 91
Bluejays, as pests in orchards and
    vineyards, 192
Bollweevil, 161
Border crops. *See* Barrier crops
Branch roots. *See* Lateral roots
Brassica spp., classification of, 23
Breeding, of plants for disease
    resistance, 78
Bromoxynil, as an herbicide, 178
Buds, 44, 45
Bud and leaf nematodes, as being
    migratory and endoparasitic, 123
Bulbs, 45–47
Bulbul, rose-whiskered, 195
Burning, of vegetation to control weeds,
    171
Burrowing nematode, as being
    migratory and endoparasitic, 123

Cabbage, control of club root in, 83
Calyx, defined, 34
Cambium, location of, 28
Canker, as a necrotic symptom of plant
    disease, 91
Captan, 116, 118
Carbamates, as herbicides, 178
Carbon monoxide, as a fumigant to
    control vermin, 189
Carboxin, 116, 118
Careers, in plant-pest control, 228
Carrying capacity
    defined, 63
    in relation to populations of harmful
        insects, 152
Carthamus tinctorius, classification of,
    23

Catabolic reactions, defined, 40
Catch crops
    in nematode control, 133–134
    in plant-disease control, 110
Ceratitus capitata, restricted to regions
    of warm climate, 154
Certification of plant-propagative parts
    to control diseases, 112
Chemosterilants, to control vermin, 191
Cherry, classification of the plant, 23
Chinch bug, creosote-treated barriers to
    protect corn from, 161
Chloraneb, 116
Chlorophacinone, anti-coagulant to
    control rodents, 201
Chlorosis, as a hypoplastic symptom of
    plant disease, 91
Chopping, of vegetation to control
    weeds, 170–171
Chrysanthemum
    classification of, 23
    damage to by leafminers, 147
Citric-acid cycle, 42
Citrus, effects of the sting nematode on,
    136
Citrus nematode, as a sedentary
    endoparasite, 123
Classification of plants, 23
Classification types, artificial and
    natural, 21
Clover, red, classification of, 23
Club root, of cabbage, control by liming
    the soil, 83
Coconut palm, classification of, 23
Cocos nucifera, classification of, 23
Coleoptera, characterized as an order of
    insects, 141
Collembola, characterized as an order of
    insects, 141
Colorado potato beetle, 155
Commensalism, defined, 66
Communities, defined, 13
Competition, 65, 66
Complex systems
    counterintuitive elements in, 210
    defined, 210
    flexibility of, 211
    undesirable characteristics of,
        210–211
Conditioning, of birds to frightening
    devices, 193–194
Contact herbicides, 180
Control
    defined in terms of plant pests, 73
    as an eradicative principle in weed
        control, 168
Copper
    as a plant-protective fungicide and
        bactericide, 116
    as the toxicant in protective
        fungicides, 118
Corm, 46, 47
Corn
    classification of, 23
    control of sting nematodes in fields of,
        132
    damage to roots of by *Trichodorus*,
        128

earworm, 145
leaf aphid, vertical resistance to, 77
seed of, 36
seedling blight, controlled by
avoidance, 81
Corolla, defined, 34
Corynebacterium, characteristics of, 25
Cotton
avoidance of bollweevils and pink
bollworms, 161
classification of the plant, 23
pink bollworms controlled by
sanitation, 157
resistance of varieties to nematodes,
129
weed control in fields by flaming, 171
Cotyledons, defined, 34
Crop losses
to diseases of plants, 100–101
to harmful insects, 8
to plant-parasitic nematodes, 122–124
to plant pests, 7–9
to vertebrate pests, 8, 188
to weeds, 166
Crop plants
defined, 4
hazards to, 4, 6
list of important, 4
relationships with pests and the
environment, 208
Crotalaria spectabilis, as a catch crop to
control nematodes, 134
Crown gall, symptoms of, 94
Cucumis melo, classification of, 23
Cucurbita spp., classification of, 23
Cultivation, to control weeds, 170
Cultural practices
to control nematodes, 133
to control plant diseases, 103, 110–115
as managerial practices to control
weeds, 171–174
Cutworms, controlled by staking and
mulching tomato plants, 157
Cybernetics, defined, 218
Cyclohexamide, as a fungicidal
antibiotic, 116
Cyst nematode, as a sedentary
endoparasite, 123
Cysts of nematodes
hatching of eggs in, 128
formation of, 130

Dagger nematode, as an ectoparasite of
plants, 123
D-D mixture, as a soil-treatment
nematicide, 116
Death rates, 61
Deer, protecting crops from, 202
Dermaptera, characterized as an order of
insects, 141
Dexon, 116
Diagnosis, of plant diseases, 98–99
Dibromochloropane, 132
Dibromochloropropane, 137
Dichlobenil, as an herbicide, 178
Dichlone, 116
Dichloropropene, as a nematicide, 133
Differentiation, of plant cells, 53

Digestion, in seeds, defined, 48
Dinocap
to control powdery mildews, 116
as a protective fungicide, 118
Dinoseb, as an herbicide, 181
Diptera, characterized as an order of
insects, 141
Discrete resistance. *See* Vertical
resistance
Disease in plants
characterized and defined, 90–91
toxins produced by parasites and, 97
Disease-progress curves, 101–102
Diseases of plants
crop losses due to, 8, 100–101
kinds of, 90
Dispersal, of inocula, 111–113
Distress calls, broadcast to frighten
birds, 194
Ditylenchus
legal restrictions against distribution
of, 130
as a migratory, endoparasitic
nematode, 123
Diuron, as an herbicide, 178
Dogs, of the genus *Cyon*, banned from
the United States, 200
Dormancy, 48
Dolichodorus, as an ectoparasitic
nematode, 123
DRC-736 and DRC-1327, to frighten
birds, 189
DRC-3324, to repel birds, 199
Dutch elm disease
control of, 83
illustrated, 93
Dwarfing, as a hypoplastic symptom of
plant disease, 91
Dynamics
defined, 60
of populations, 60–61

Early blight of potato, simulated
epidemic of, 220–221
Ecology, defined, 11, 13
Economic threshold, defined, 72
Ecosystems, defined, 13
Ectoparasitic nematodes, defined, 128
Education
distinguished from training, 225
as an example of a system, 213
in plant-pest control, 225
Elm, wilt (Dutch elm disease) of, 93
Elongation, region of in roots, 50, 52
Embryo, 34
Endocarp, 28
Endodermis, 28
Endoparasitic nematodes, defined, 128
Endosperm, 34
Energy, flow through ecosystems,
16–17, 140
Endrin, to protect pine cones from birds,
189–190
Entomology, definition of applied, 139
Ephemeroptera, characterized as an
order of insects, 141
Epicotyl, 34

EPIDEM, a computer-simulation of a plant disease, 220
EPIDEMIC, a computer-simulation of a plant disease, 221
Epidemics of plant diseases, 97, 98
Epidemiology, defined, 97, 98
EPISEM, a computer-simulation of a plant disease, 221
Eradication
    as a principle of insect control, 157–158
    as a principle of plant-disease control, 119
    as a principle of plant-pest control, 75
    as a principle of vermin control, 195–198
    as a principle of weed control, 168–169
Erwinia, characteristics of, 25
Ethoprop, as a nematicide, 132
Ethylene dibromide, 132
Exclusion
    as a principle of insect control, 156–157
    as a principle of plant-disease control, 119
    as a principle of plant-pest control, 74–75
    as a principle of vermin control, 195
    as a principle of weed control, 167
Exploders, to frighten bird pests, 193

Facultative parasites, defined, 67
Facultative saprophytes, defined, 67
Fallowing, to control nematodes, 135
Federal Bird Acts, 195
Federal Environmental Pesticide Control Act (FEPCA), 230
Federal Insecticide Act, 230
Federal Insecticide, Fungicide, and Rodenticide Act (FIFRA) and amendments thereto, 230
Federal Noxious Weed Act, 168
Federal Seed Act, 168
Feedback loops of information
    defined, 215
    illustrated, 217
    variables in, 216
Foodstuffs, storage of in plants, 43
Feltia subterranea, control of in tomato fields, 157
Fensulfothion, as a nematicide, 132–133
FEPCA, 230
Ferbam, as a protective fungicide, 118
"Fertile Crescent," 4, 5
Fibrous-root systems, defined, 51
Field immunity. *See* Vertical resistance
Field resistance. *See* Horizontal resistance
FIFRA, 230
Filament, of stamens, 34
Flaming, to control weeds, 171
Flea beetles, 157
Flooding of soil
    to control nematodes, 135
    to control weeds, 171
    to control wireworms, 157
Flow variables, defined, 216
Flowers, characterized, 33

Fringe crops. *See* Barrier crops
Fruit, parts of, 34
Function, as an element of a system, 209
Functions of plants
    as adversely affected by diseases, 90
    as the basis for classifying plant diseases, 39
    vital, 39
Fundamental research, as distinguished from applied research, 227
Fumarin, to control vermin, 201
Fungicides, exemplifying principles of plant-disease control, 115

Galls
    as hyperplastic symptoms of plant disease, 91
    meristematic activities in, 54
Gametes, 37
Gametophytic generations, 38
Gas-bomb exploders, 193
Geese, used to control weeds in beet fields, 195
"Giant cells," in plant tissue parasitized by nematodes, 126
Glycine max, classification of, 23
Glycolysis, 41
Grasshopper, as an orthopterous insect, 141
Gossypium hirsutum
    classification of, 23
    controlling weeds in fields of, 171
Grazing, to control weeds, 174
Growth
    control of pests affecting, 82–83
    primary and secondary in plant stems and roots, 29–31
Gymnospermae, characterized as a class of plants, 26
Gypsy moth, as a phytophagous insect, 140

Hand pulling, of weeds, 169
"Hatching factors," for eggs of cyst-forming nematodes, 128
Helicotylenchus, 124
    as a migratory, semi-endoparasitic nematode, 123
Helminthosporium victoriae, as the cause of Victoria blight of oats, 9
Hemicycliophora, as a migratory, semi-endoparasitic nematode, 123
Hemiptera, characterized as an order of insects, 141
Herbicides
    application of, 180–184
    defined and classified, 177
    formulations of, 178
    selectivity of, 178–180
Herbivores, vertebrate pests as, 68
Hessian fly
    avoided by late planting of wheat, 85, 161
    vertical resistance of wheat varieties to, 77
Heterodera
    schachtii, controlled by crop rotation, 130
    as a sedentary, endoparasitic nematode, 123

Meristems
  defined, 29
  locations of, 29, 30
  pest damage to, 53, 54
  in roots, 50, 52
  tolerance to parasites, 54
Mesocarp, 34
Mestranol
  to control rodents, 201
  to protect pine cones from birds, 190
Metamorphosis of insects, 149
Metabolism, defined, 40
Metham, as an herbicide, 178
Methyl bromide
  as a general pesticide for soil
    fumigation, 116, 117
  as a fumigant for vermin control, 189
Mexican bean beetle, 144
Migration, of plant pests, 61, 62
Migratory Bird Act, 195
MIT ("Vorlex"), as a soil-treatment
    fungicide and nematicide, 116
Mites
  families of, 148
  host-plants of, 148
  mouthparts of, 145, 146
Mitosis, 37
MLO's. See Mycoplasmalike organisms
Moles, control by trapping, 202
Molting
  of insects, 150
  of nematodes, 127
Mongoose, 200
Monogenic resistance. See Vertical
    resistance
Monuron, as an herbicide, 178
Morphology
  defined, 21
  diversity of among plants, 26
Mortality, 61
Mosaic, as a hypoplastic symptom of
    plant disease, 91
Mosaic viruses, spread of not checked
    by insecticides, 84, 210–211
Mouthparts, of insects
  chewing types, 142, 143
  piercing-sucking types, 142, 143, 144
  rasping types, 144
  sponging-lapping types, 144
Mouthparts, of mites, 145–146
Mowing, to control weeds, 170–171
Mulches
  of aluminum to repel aphids, 157
  of green manure as being toxic to
    nematodes, 135
Musa sapientum, classification of, 23
Muskmelon, classification of, 23
Muskrats, control by trapping, 202
Mutualism
  defined, 53
  as an interaction between two
    populations, 66
Mycoplasmalike organisms
  control of, 84
  as infectious agents of plant diseases,
    95
Mycorrhizae, 53

Natality, 61
National Environmental Policy Act, 230

Natural control of insects, 153–154
Necrosis, as a class of plant-disease
    symptoms, 91
Needle nematode, as an ectoparasite of
    plants, 123
Nematicides, kinds of, 132–133
Nematodes
  characteristics of, 122–126
  control of root-infecting, 83
  life cycles of, 127–129
  numbers required to induce disease,
    131
  stimulatory effects on plant hosts, 131
Neuroptera, characterized as an order of
    insects, 141
Neutralism, 66
Nitrifying bacteria, harmed by
    dichloropropene, 133
Nomenclature,
  Codes of, 21, 22
  defined, 22
  procedures used in, 22, 23
Noninfectious agents, of plant diseases,
    94
Noninfectious diseases, of plants, 94
Numerical taxonomy, 24
Nutria, controlled by trapping, 202
Nutsedge, in a field of peanuts, 165

Oats
  rust-resistance in varieties of, 9
  seedling of, 49
  susceptible to Victoria blight, 9
Obligate parasites, defined, 69
Odonata, characterized as an order of
    insects, 141
Offshoots, kinds of, 45
Organs of plants, 26
Orthoptera, characterized as an order of
    insects, 141
Oryza sativa, classification of, 23
Ovary, 35
Ovule, 34
Oxathiin, 116
Oxidative respiration, 42

Pangola grass, toxic to nematodes,
    133–134
Paraquat, as an herbicide, 180
Parasites
  kinds of, 67
  in insect control, 154
  life cycles of plant parasites, 95, 97
  in vermin control, 190–191
  in weed control, 175–176
Parasitism, 66
Paratylenchus, as a migratory,
    semi-endoparasitic nematode, 123
Pathogens, of plants, 11
PCNB, as a soil-treatment fungicide, 116
Peaches, weed control in orchard of, 174
Peanuts
  continuous cropping and effect on
    nematodes of, 134
  in herbicide-treated fields, 165, 179,
    183
  weed control in fields of, 165, 179, 183
Pedicel, 34
Peduncle, 34

Quarantines *(continued)*
  to control plant parasites, 112
  as an exclusionary measure of
    plant-pest control, 75
Quelea quelea, 188

Rabbits
  controlled by the virus of
    myxomatosis, 191
  protection of crops from, 202
Radiation, to induce sterility in vermin,
    191
Radopholus
  legal restrictions against the spread of,
    131
  as a migratory endoparasitic
    nematode, 123
  similis, persistence in soil, 128
Rate variables, in feedback loops, 216
Rats
  crop and grain losses to, 8
  multimammate kinds banned from the
    United States, 200
Red-ring nematode, migratory and
  endoparasitic, 123
Redwing blackbirds, controlled by
  DRC-1327, 189
Refuges, for wildlife to control vermin,
    192
Reniform nematode, sedentary and
  endoparasitic, 123
Repellents, to control birds, 198–199
Research, kinds and procedures of, 227
Resistance
  of insects to insecticides, 158
  of pests to pesticides, 76–77
  of plants to diseases
    horizontal, 109
    practical problems associated with,
      108–109
    selection of varieties for, 105
    sources of resistance, 104
    vertical, 109
    See also Horizontal resistance,
      Vertical resistance
  of plants to nematodes, 135–136
  of vermin to anticoagulants, 189
Respiration, defined and described,
    40–43
Rest, defined, 48
Retarding factors, 15, 16
Rhadinaphelenchus, as a migratory,
  endoparasitic nematode, 123
Rhizobium, as an infectious bacterium
  beneficial to its hosts, 51, 53
Rhizomes, 45
Rice, classification of the plant, 23
Rickettsiae, as infectious agents of plant
  diseases, 95
Robins, as pests in orchards and
  vineyards, 192
Root cap, 50, 52
Root hairs, 51, 52
Root knot, of tomato, 127
Root-knot nematode, sedentary and
  endoparasitic, 123
Root-lesion nematode, migratory and
  endoparasitic, 123

Roots
  as affected by plant pests, 44
  diagram of, 31
  food storage in, 43, 44
  functions of, 43
  lateral, 51, 52
  specialized, 43, 44
  systems of, 51
  tissues of, 30, 31, 52
Rosa spp., classification of, 23
Rose
  blackspot disease of, 92
  powdery mildew of, 96
Rosetting, as a hypoplastic symptom of
  plant disease, 91
Rosy pastor, 195
Rot, as a necrotic symptom of plant
  disease, 91
Rotation of crops
  to control insects, 157–158
  to control nematodes, 129–130
  to control plant diseases, 110
  to control weeds, 172
Rotylenchulus
  reniformis, in tobacco roots, 127
  as a sedentary, endoparasitic
    nematode, 123
Rotylenchus, as a migratory,
  endoparasitic nematode, 123
"*r* strategists," 152
Runners, as specialized stems, 45, 46

Saccharum officinarum, classification
    of, 23
Safflower, classification of, 23
St. John's wort, controlled by goatweed
    beetles, 175
Sanitation, to control insects, 157
Saprophytes, facultative, 67
Scientific method, described, 227
Screening of closed structures
  to protect plant products from birds, 199
  to protect plants from insects, 161
Scutellonema, as a migratory,
    semi-endoparasitic nematode, 123
Secondary cycles, of plant parasites, 97
Seed coat, 34
Seed-gall nematode, sedentary and
  endoparasitic, 123
Seedlings
  of bean, 49
  blights of controlled by avoidance, 81
  control of pests affecting, 82
  defined, 49, 50
  of oats, 49
  pest damage to, 50
Seed plants, life cycles of, 38
Seeds
  food storage in, 47
  germination of, 48, 49
  parts of, 34
Seed-treatment fungicides, 115–116
Selective herbicides
  action of, 179–180
  defined, 178
Sepals, 34
Sheath nematode, migratory and
  semi-endoparasitic, 123

240

Shelford's law of toleration, 13–14
Shot-hole borers, 157
Shrapnel bombs, to control birds, 193
Signs, as structures of plant parasites, 99
Simazine, as an herbicide, 178
Siphonaptera, characterized as an order
    of insects, 141
Sirens, to frighten birds, 193
Slips, as offshoots of stems, 45
SMDC (vapam), 116
Smuts, control of in cereal grains, 83
Sodium fluoroacetate (1080)
    as an animal poison, 189
    to control mammalian pests of plants,
    201
Soil-borne plant parasites, control of,
    110–111
Soil treatments
    fungicides as, 117
    insect control by, 157
Solar energy, use of, 17, 18
Sorghum, brown seed of, 194
Soybean
    avoidance of velvet-bean caterpillar
        by, 161
    classification of, 23
Soybean cyst nematode, failure of
    quarantine measures to control, 131
Species, defined, 12
Sphaerotheca pannosa, life cycle of, 96
Spiral nematode, migratory and
    semi-endoparasitic, 123
Sporophytic generations, 38
Spot, as a necrotic symptom of disease,
    91
Spring canker worms, 160–161
Spruce budworm, control of, 210
Spurs, as offshoots of stems, 45, 46
Squirrels, controlled with mestranol,
    201
Stamens, 34
Starlicide, to control birds, 196
State variables, in feedback loops, 216
Stele, 28
Stem nematode, migratory and
    endoparasitic, 123
Stems, 44–46
Stigma, of flowers, 34
Sting nematode, as an ectoparasite of
    plants, 123
Stinkbug, as a hemipterous insect, 141
Stolons, as specialized stems, 45
Stomata
    as avenues of entry for plant parasites,
    97
    form and function of, 56
Storage of food
    control of pests affecting, 82
    in plants, 43–47
    in roots, 43, 44
    in stems, 44–45
Streptomycin, 116
Strip cropping, to control insects, 157
Structure, as an element of a system, 209
Strychnine
    as an animal poison, 189
    to control mammalian pests of plants,
    201

Stubby root of corn, 128
Stubby root nematode, 123
Style, 34
Stylet nematode, as an ectoparasite of
    plants, 123
Subsystems, of plant-pest control, 224
Suckers, as offshoots of stems, 45
Sudan dioch, 195
Sugar beet, classification of, 23
Sugar cane, classification of, 23
Sulfur, as a fungicide, 116, 118
Surfactants, to control birds, 192–193
Survival, of plant parasites between
    cropping seasons, 110–111
Survivorship patterns, of different
    species, 64, 65
Symbiosis, 66
Symptoms of plant diseases, 91, 98
Synecology, defined, 13
Systemic insecticides, to control
    sap-feeding arthropods, 158
Systemic pesticides, as therapeutants,
    76
Systems
    of agriculture, 223
    boundaries of, 213
    complex, 210–211
    components of, 215
    coordination of, 227–228
    defined, 209
    elements of, 209, 212
    environment, without and within,
        213–214
    management of, 215
    nature of, 208–209
    of plant-pest control, 223, 224
    resources of, 214
    structure of, 212
Systems analysis, 218–219
Systems approach
    application to plant-pest control, 219
    principles of, 210–211

Tap-root systems, 51
Tariff Act, to regulate imports of bird
    feathers and skins, 195
Taxon, defined, 12
Taxonomy, 22, 24
Tendrils, as specialized stems, 45
Terminology, 11
    of weed and plant-pest control, 167
Testa, 34
Thallium sulfate, to control vermin, 189
Therapy
    to control insects, 158
    to control nematodes, 135
    to control plant diseases, 111
    as a curative pest-control principle, 76
    by heat, 111
    inapplicable to the control of vermin,
        188
    as a principle of plant-pest control,
        76–77
Thermodynamics, first and second laws
    of, 16
Thiram, as a seed-treatment fungicide,
    116
Thorns, as modified stems, 45